# Indoor Air Quality Guide

This is an ASHRAE Design Guide. Design Guides are developed under ASHRAE's Special Publication procedures and are not consensus documents. This document is an application manual that provides voluntary recommendations for consideration in achieving improved indoor air quality. This publication was developed under the auspices of ASHRAE Special Project 200.

This publication was developed under the auspices of ASHRAE Special Project 200.

Development of this Guide was funded by U.S. Environmental Protection Agency (EPA) Cooperative Agreement XA-83311201.

## INDOOR AIR QUALITY GUIDE—PROJECT COMMITTEE



## INDOOR AIR QUALITY GUIDE—STEERING COMMITTEE



**Any updates/errata to this publication will be posted on the ASHRAE Web site at www.ashrae.org/publicationupdates.**

# Indoor Air Quality Guide

## Best Practices for Design, Construction, and Commissioning

American Society of Heating, Refrigerating and Air-Conditioning Engineers
The American Institute of Architects
Building Owners and Managers Association International
Sheet Metal and Air Conditioning Contractors' National Association
U.S. Green Building Council
U.S. Environmental Protection Agency

ISBN: 978-1-933742-59-5


1791 Tullie Circle, NE
Atlanta, GA 30329
www.ashrae.org


Printed in the United States of America

Cover design by Adam Amsel.

Design and illustrations by Adam Amsel, Scott Crawford, and Katie Freels.
Design Consultant: Francis D.K. Ching

Technical Editor: David Mudarri

---



---

Library of Congress Cataloging-in-Publication Data

Indoor air quality guide : best practices for design, construction, and commissioning.
p. cm.
Includes bibliographical references.
Summary: "Comprehensive, practical resource on design and construction for enhanced indoor air quality (IAQ) in commercial and institutional buildings. Useful for architects, engineers, building owners concerned with high-quality indoor environment. Focuses on today's major IAQ issues: moisture management, ventilation, filtration and air cleaning, and source control"--Provided by publisher.
ISBN 978-1-933742-59-5 (pbk.)
1. Commercial buildings. 2. Public buildings.. 3. Buildings--Environmental engineering. 4. Indoor air quality. 5. Indoor air pollution. I. American Society of Heating, Refrigerating and Air-Conditioning Engineers.
TH6031.I53 2009
613'.5--dc22

2009017909

---

# CONTENTS

## PART II—Detailed Guidance

## Detailed Information for Design, Construction, and Commissioning for IAQ

*Detailed guidance for each Strategy is available on the CD accompanying this book.*

## Acknowledgments

The *Indoor Air Quality Guide* provides a long-needed resource to the building community. The project committee (PC) members who wrote the Guide worked extremely hard to pull together practical, technically sound information covering the full range of indoor air quality (IAQ) issues important to practitioners. The authors brought to this project many years of experience in design, construction, commissioning (Cx), forensic IAQ investigation, research, and facilities management. They met 14 times in 20 months and held numerous conference calls. Between meetings, they critically reviewed and digested IAQ research and practical experience, reviewed each others' drafts, solicited input from expert colleagues, weighed reviewer input, and worked very hard to make the information accessible and useful for designers, contractors, architects, owners, operators, and others. In collaboration with ASHRAE's Chapter Technology Transfer Committee, they also developed and presented a satellite broadcast/webcast based on the Guide in April 2009. We want to extend our personal thanks and appreciation to this group for the extraordinary effort they poured into this Guide and for the passion they displayed for providing quality indoor environments for building occupants.

The PC's efforts were guided by a steering committee (SC) made up of members from all of the cooperating organizations: American Institute of Architects (AIA), American Society of Heating, Refrigerating and Air-Conditioning Engineers (ASHRAE), Building Owners and Managers Association (BOMA) International, U.S. Environmental Protection Agency (EPA), Sheet Metal and Air-Conditioning Contractors' National Association (SMACNA), and U.S. Green Building Council (USGBC). The SC assembled an expert team of authors and defined a project scope that kept the PC's task manageable and focused. The representatives from these organizations brought a collegial and constructive spirit to the task of setting policy for the Guide.

This project would not have been possible without the funding provided by the EPA through Cooperative Agreement XA-83311201 with ASHRAE. In recognizing the depth of expertise available in the building community and its unique ability to educate its own members, EPA helped to make the Guide accurate, practical, and credible. ASHRAE's Washington, D.C., office had the vision to recognize the EPA solicitation as an opportunity to develop guidance that had long been needed and was a longtime goal of ASHRAE's. Bruce Hunn and other ASHRAE staff developed ASHRAE's response to the solicitation and executed the cooperative agreement with EPA.

Several people not on the PC or SC played key roles in bringing this Guide to fruition. Adam Amsel, Scott Crawford, and Katie Freels, master's-level graduates of the University of Washington, Seattle, produced the illustrations, created the cover, and designed the layout for all of Part I and Part II. Francis D.K. Ching of the University of Washington served as a design consultant in this effort. Their team did a wonderful job in making the Guide attractive to architects while preserving the nuts and bolts functionality desired by engineers, in spite of many schedule changes and other challenges. David Mudarri, PhD, joined the project as technical editor and helped make the Guide much clearer and more coherent.

Two focus groups provided valuable early input that helped to shape this Guide. Sixty people participated in the two peer reviews, providing 1200 remarks that helped to strengthen and clarify the Guide. We appreciate the considerable time the focus group participants and reviewers took from their busy schedules to give us their thoughtful input and hope that they see the impacts of their recommendations in the finished product.

Individual PC members sought input from other experts to ensure that they captured the latest and best thinking on their topics, and we appreciate those contributions. A number of people generously provided case studies for inclusion in the Guide; their names are provided in the case study credits. Mary Sue Lobenstein of Minneapolis coordinated the production and revision of all graphics included in the Guide. She also took charge of Strategy 3.6 on pest control when the PC author was unable to complete the work. Michael Merchant, PhD, an entomologist at Texas A&M University, graciously provided considerable assistance in "getting the bugs out" of this Strategy.

Lilas Pratt of ASHRAE staff went beyond the call of duty in managing an enormous number of documents and sometimes prickly authors with great competence and efficiency. Cindy Michaels, Managing Editor of ASHRAE Special Publications, did a tremendous job making the document into a first-class publication. Their efforts, as well as those of many other ASHRAE staff persons, are greatly appreciated.

Notwithstanding all of the help generously given by many people, the PC remains responsible for the contents of the Guide. We hope you find it useful.

Martha J. Hewett
*Chair, Special Project 200*

Andrew Persily
*Chair, IAQ Guide Steering Committee*

October 2009

## Abbreviations and Acronyms

AABC = Associated Air Balance Council
AAMA = American Architectural Manufacturers Association
ACGIH = American Conference of Governmental Industrial Hygienists
ADC = Air Diffusion Council
AHAM = American Home Appliance Manufacturers
AHU = air-handling unit
AIA = American Institute of Architects
AIHA = American Industrial Hygiene Association
AMCA = Air Movement and Control Association
ANSI = American National Standards Institute
AQS = Air Quality Sciences, Inc.
ASD = active soil depressurization
ASHE = American Society of Hospital Engineers
ASHRAE = American Society of Heating, Refrigerating and Air-Conditioning Engineers
ASTM = ASTM International (formerly the American Society for Testing and Materials)
BIFMA = Business and Institutional Furniture Manufacturer's Association
BoD = Basis of Design
BOMA = Building Owners and Managers Association International
CARB = California Air Resources Board
CDC = Centers for Disease Control and Prevention
CDHS = California Department of Health Services
CEC = California Energy Commission
CHPS = Collaborative for High Performance Schools
CO = carbon monoxide
$CO_2$ = carbon dioxide
CoC = contaminants of concern
CREL = Chronic Reference Exposure Level
CV = constant volume
Cx = commissioning
CxA = commissioning authority
DCV = demand-controlled ventilation
DDDF = dual duct dual fan
DOAS = dedicated outdoor air system
DOE = U.S. Department of Energy
DX = direct expansion
EDR = Energy Design Resources
EPA = U.S. Environmental Protection Agency
ERV = energy recovery ventilator
EU = European Union
FAC = filtration and gas-phase air cleaning
FEMA = Federal Emergency Management Agency
HE = high efficiency
HEPA = high-efficiency particulate air
HVAC = heating, ventilating, and air conditioning
IAQ = indoor air quality
IAQP = IAQ Procedure
IEQ = indoor environmental quality
IEST = Institute of Environmental Sciences and Technologies
IPCC = Intergovernmental Panel on Climate Change
ITRC = Interstate Technology & Regulatory Council

JCAHO = Joint Commission on Accreditation of Healthcare Organizations
MDF = medium density fibreboard
ME = medium efficiency
MERV = Minimum Efficiency Reporting Value
NAAQS = National Ambient Air Quality Standards
NAIMA = North America Insulation Manufacturers Association
NASA = National Aeronautics and Space Administration
NCARB = National Council of Architectural Registration Boards
NEBB = National Environmental Balancing Bureau
NEHA NRPP = National Environmental Health Association National Radon Proficiency Program
NFPA = National Fire Protection Association
NFRC = National Fenestration Rating Council
NIBS = National Institute of Building Sciences
NIH = National Institutes of Health
NIOSH = National Institute of Occupational Safety and Health
$NO_2$ = nitrogen dioxide
$NO_x$ = nitrogen oxides
NRCan = Natural Resources Canada
NRC-IRC = National Research Council Canada Institute for Research in Construction
NRSB = National Radon Safety Board
O&M = operation and maintenance
OPR = Owner's Project Requirements
OSB = oriented strand board
OSHA = Occupational Safety and Health Administration
OTA = U.S. Congress Office of Technology Assessment
PM10 = particulate matter with a diameter of 10 μm or less
PM2.5 = particulate matter with a diameter of 2.5 μm or less
PVC = polyvinyl chloride
QA = quality assurance
RH = relative humidity
RTU = rooftop unit
SMACNA = Sheet Metal and Air Conditioning Contractors' National Association
$SO_2$ = sulfur dioxide
SVOC = semi-volatile organic compound
TAB = testing, adjusting, and balancing
TABB = Testing, Adjusting, and Balancing Bureau
TLV = threshold limit value
TVOC = total volatile organic compound
TWA = time-weighted average
UL = Underwriters Laboratories
USGBC = U.S. Green Building Council
UVGI = ultraviolet germicidal irradiation
VAV = variable-air-volume
VFD = variable-frequency drive
VOC = volatile organic compound
VRP = Ventilation Rate Procedure
WHO = World Health Organization

## Foreword: Why this Guide Was Written

Buildings are expected to fulfill a variety of requirements related to their function, applicable codes and standards, and environmental and community impacts. Among these requirements, indoor air quality (IAQ) is typically addressed through compliance with only minimum code requirements, which are based on industry consensus standards such as *ANSI/ASHRAE Standard 62.1, Ventilation for Acceptable Indoor Air Quality* (ASHRAE 2007a). Yet IAQ affects occupant health, comfort, and productivity, and in some cases even building usability, all of which can have significant economic impacts for building owners and occupants.

While building owners and building professionals may recognize the importance of IAQ, they often do not appreciate how routine design and construction decisions can result in IAQ problems. In addition, they may assume that achieving a high level of IAQ is associated with premium costs and novel or even risky technical solutions. In other cases, they may employ individual measures thought to provide good IAQ, such as increased outdoor air ventilation rates or specification of lower emitting materials, without a sound understanding of the project-specific impacts of these measures or a systematic assessment of IAQ priorities.

Information exists to achieve good IAQ without incurring excessive costs or employing practices that are beyond the current capabilities of the building professions and trades. This Guide, resulting from a collaborative effort of six leading organizations in the building community[1] and written by a committee of some of the most experienced individuals in the field of IAQ, presents best practices for design, construction, and commissioning (Cx) that have proven successful in other building projects. It provides information and tools architects and design engineers can use to achieve an IAQ-sensitive building that integrates IAQ into the design and construction process along with other design goals, budget constraints, and functional requirements. While some key issues in the field of IAQ remain unresolved, this document presents the best available information to allow practitioners to make informed decisions for their building projects.

The Guide addresses the commercial and institutional buildings covered by ASHRAE Standard 62.1 and was written with the following audiences in mind:

- Architects, design engineers, and construction contractors who can apply the recommended practices during design and construction processes.
- Building owners, developers, and other decision makers who can use this Guide to direct the work of these professionals.
- Commissioning authorities (CxAs) who can ensure that design elements, construction schedules, construction observation, and functional testing are appropriate to meet IAQ-related goals and requirements.
- Product and material specifiers for both new and existing buildings who can choose materials and products with lower IAQ impact.
- Organizations that provide sustainable building rating programs and/or that conduct training for these programs.
- Facility managers and building operators who may use the Guide to understand the IAQ implications of existing systems and operations and maintenance practices.

[1] The six organizations contributing to the creation of this Guide are American Society of Heating, Refrigerating and Air-Conditioning Engineers (ASHRAE); American Institute of Architects (AIA); Building Owners and Managers Association (BOMA) International; U.S. Environmental Protection Agency (EPA); Sheet Metal and Air Conditioning Contractors' National Association (SMACNA); and U.S. Green Building Council (USGBC).

## Message to Building Owners

Indoor air quality (IAQ) is one of many issues that building owners and developers must address to provide buildings that meet their needs and the needs of the building occupants. While building occupants do sometimes complain about poor IAQ, it is not always on the top of their list of concerns. So why should you worry about IAQ when you have so much else to worry about?

- First, better IAQ leads to more productive and happier occupants. In commercial real estate, satisfied occupants are tied directly to return on investment and bottom-line economics, while in schools and institutional buildings they are tied to learning outcomes and organizational missions. While it is hard to put firm numbers on these benefits, there is increasing evidence of measurable productivity increases and reduced absentee rates in spaces with better IAQ.[1] In considering the economics of IAQ, it is important to note that the salaries of building occupants are the largest cost associated with building operation, dwarfing energy by a factor of 50 or even 100.
- Second, IAQ problems that get out of hand can be quite costly in terms of lost work time, lost use of buildings, expensive building or mechanical system repairs, legal costs, and bad publicity. While extreme IAQ problems are rare, they do occur, and the consequences can be dramatic. Less severe problems are more common and can erode occupant productivity, affect occupancy and/or rent levels, and lead to costs for smaller legal disputes or repairs.

This document presents a wealth of practical information on how to design and construct buildings with better IAQ without large financial investments or untested technologies. While the Guide is full of information on design and construction to control moisture, reduce contaminant entry, and provide effective ventilation, probably the most important message for the owner/developer is to put IAQ on the table at the very beginning of the development and design processes. Including IAQ in the earliest discussions with the architect and the rest of the project team will make it easier and more effective to provide good IAQ at lower or even no added cost.

By the time a building's schematic design is complete, many opportunities to achieve good IAQ have been foreclosed, which can easily result in unintended consequences or expensive and inadequate "force fitting" of solutions. When IAQ, energy efficiency, and other project objectives are considered together at the initial design phases, design elements for each objective can be mutually reinforcing rather than at odds with one another.

[1] For a good overview of research quantifying IAQ health and productivity impacts, visit the IAQ Scientific Findings Resource Bank at http://eetd.lbl.gov/ied/sfrb/sfrb.html. The IAQ-SFRB is jointly administered by the U.S. Environmental Protection Agency and Lawrence Berkeley National Laboratories.

# Introduction

### Why Good IAQ Makes Sense

Indoor air quality (IAQ) is one of many factors that determine building functionality and economics. IAQ affects building occupants and their ability to conduct their activities; creates positive or negative impressions on customers, clients, and other visitors to a building; and can impact the ability to rent building space. When IAQ is bad, building owners and managers can find themselves devoting considerable resources to resolving occupant complaints or dealing with extended periods of building closure, major repair costs, and expensive legal actions. When IAQ is good, buildings are more desirable places to work, to learn, to conduct business, and to rent.

IAQ directly affects occupant health, comfort and productivity. Well-established, serious health impacts resulting from poor IAQ include Legionnaires' Disease, lung cancer from radon exposure, and carbon monoxide (CO) poisoning. More widespread health impacts include increased allergy and asthma from exposure to indoor pollutants (particularly those associated with building dampness and mold), colds and other infectious diseases that are transmitted through the air, and "sick building syndrome" symptoms due to elevated indoor pollutant levels as well as other indoor environmental conditions. These more widespread impacts have the potential to affect large numbers of building occupants and are associated with significant costs due to health-care expenses, sick leave, and lost productivity. The potential reductions in health costs and absenteeism and improvements in work performance from providing better IAQ in nonindustrial workplaces in the U.S. are estimated to be in the high "tens of billions of dollars annually" (EPA 1989; Fisk 2000; Mendell et al. 2002).[1]

Despite these significant impacts, many building design and construction decisions are made without an understanding of the potentially serious consequences of poor IAQ and without benefit of the well-established body of knowledge on how to avoid IAQ problems. While controlling indoor pollutant levels and providing adequate ventilation and thermal comfort have motivated the design and use of buildings for centuries, awareness of and concerns about IAQ have increased in recent decades. However, in most cases IAQ is still not a high-priority design or building management concern compared to function, cost, space, aesthetics, and other attributes such as location and parking.

Given the very real benefits of good IAQ and the potentially serious consequences of poor IAQ, building owners, designers, and contractors can all benefit from an increased focus on providing good IAQ in their buildings. This Guide can enhance all parties' ability to design, construct, and operate buildings with good IAQ using proven strategies that do not incur significant additional costs.

#### The High Cost of Poor IAQ

The costs of poor IAQ can be striking. There have been many lawsuits associated with IAQ problems, though most are settled with no financial details released. However, some publicly disclosed cases have involved legal fees and settlements exceeding $10 million. For example:

- In 1995, Polk County, Florida, recovered $47.8 million in settlements against companies involved in the construction of the county courthouse (including $35 million from the general contractor's insurer), due to moisture and mold associated with building envelope problems. The original construction cost for the building was $35 million, but $45 million was spent to replace the entire building envelope, clean up the mold, and relocate the court system.
- Occupants of a courthouse in Suffolk County, Massachusetts, received a $3 million settlement in 1999 following a series of IAQ problems associated with a combination of inadequate ventilation and fumes from a waterproofing material applied to the occupied building.

Numerous IAQ problems have also occurred in private-sector buildings, but these tend to be settled out of court and are therefore not in the public record. As in public buildings, the causes of the problems vary and the settlement costs can be very expensive. A conservative estimate puts the lower bound of litigation costs during the early 2000s well over $500 million annually.

### What is Good IAQ?

This Guide is intended to help architects, contractors, and building owners and operators move beyond current practice to provide "good IAQ." Good IAQ is achieved by providing air in occupied spaces in which there are no known or expected contaminants at concentrations likely to be harmful and no conditions that are likely to be associated with occupant health or comfort complaints and air with which virtually no occupants express dissatisfaction. It includes consideration of both indoor air pollution levels and thermal environmental parameters. However, the limits

[1] Other research related to impacts of IAQ is available at the IAQ-SFRB at http://eetd.lbl.gov/ied/sfrb/sfrb.html.

of existing knowledge regarding the health and comfort impacts of specific contaminants and contaminant mixtures in nonindustrial environments, coupled with the variations in human susceptibility, make it impossible to develop a single IAQ metric that can provide a summary measure of IAQ in buildings.

In the context of this Guide, then, good IAQ results from diligent compliance with both the letter and intent of ASHRAE Standard 62.1 (ASHRAE 2007a), technically sound and well-executed efforts to meet or exceed these minimum requirements, and the application of IAQ-sensitive practices in building and system design, construction, commissioning (Cx), and operation and maintenance (O&M) throughout the life of a building. It is reasonable to assume that adherence to today's minimum standards, i.e., ASHRAE Standard 62.1, and to good engineering and O&M practices will result in acceptable IAQ. However, current practice does not always achieve compliance with minimum standards or with good practice, and many building owners and practitioners desire to achieve better-than-acceptable IAQ. These are the primary motivations for the development of this Guide.

**Importance of the Design and Construction Process**

While there is ample information and experience on achieving good IAQ in commercial and institutional buildings, it doesn't happen automatically. It takes a level of awareness and commitment that isn't typical of most projects, including an effort to make IAQ part of the design at the very beginning of the project. There are two primary reasons to include IAQ considerations in the earliest stages of project planning: avoiding problems that occur when IAQ is treated as an afterthought and allowing consideration of alternative design concepts that involve decisions made early in the design process.

Incorporating IAQ at the very beginning of conceptual design gets a number of key issues before the design team, enabling them to make informed decisions that will affect the project through the construction and occupancy phases. These issues and decisions are addressed in more detail in this document but include the owner's expectations for IAQ in the building, outdoor contaminant sources in or near the site, the activities expected to occur in the building (and the contaminants that might be associated with these activities), the characteristics of the occupants (e.g., their age range and health status, as well as the possibility of short term visitors that may have very different expectations than occupants who will remain in the building for a long time), and the approaches used to heat, cool and ventilate the building. If these considerations are not addressed until after the building layout is defined, the ventilation system type is selected, and the ventilation rate design calculations are complete, it will be difficult if not impossible to accommodate the particular needs of the building, its owner, and its occupants.

Many design decisions that can lead to poor IAQ are made in the early phases of design and are difficult to modify or correct later on. Early design missteps can be avoided if IAQ is put on the table as a key design issue at the start. Examples are inadequate space for mechanical equipment, limiting access for inspection and maintenance, and selection of interior finishes that can lead to high levels of volatile organic compound (VOC) emissions or to moisture problems in the building envelope.

Making IAQ part of the initial discussion of design goals—on par with building function, image, and energy use—allows consideration of high-performance design concepts that can support good IAQ, energy efficiency, and other important design goals. Examples include mechanical systems that separate outdoor air ventilation from space conditioning, the application of natural ventilation, high-efficiency air cleaning in conjunction with lowered ventilation rates, and the selection of low-emitting materials based on sound technical consideration of the options.

Making a commitment to good IAQ at the beginning of a project and maintaining that focus through design, construction, and Cx will result in a building that is more successful in meeting its design goals and achieving the desired level of performance throughout its life.

**What are the IAQ Problems in Buildings?**

The information in this Guide is based on the IAQ problems that have been occurring in commercial and institutional buildings for several decades and the authors' experience in investigating, resolving, and avoiding these problems. The causes of these problems were used to develop the organization of this Guide.

### *IAQ during Design and Construction*

Many IAQ problems are the result of IAQ not being considered as a key issue at the very beginning of the design process. Basic design decisions related to site selection, building orientation, and location of outdoor air intakes and decisions on how the building will be heated, cooled, and ventilated are of critical importance to providing good IAQ. Efforts to achieve high levels of building performance without diligent considerations of IAQ at the beginning of the design process often lead to IAQ problems and represent missed opportunities to ensure good IAQ.

### *Lack of Commissioning*

While a good design is critical to providing good IAQ, if the building systems are not properly installed or commissioned so that they operate as designed, IAQ conditions may be seriously compromised. Therefore, a key factor in achieving good IAQ is a serious commitment to a comprehensive Cx effort that starts in the design phase and continues well into occupancy. This effort should include a focus on Cx of systems and assemblies critical to good IAQ.

### *Moisture in Building Assemblies*

There have been many notable cases of building IAQ problems associated with excessive levels of moisture in building assemblies, particularly in the building envelope. Such situations can lead to mold growth that can be very difficult to fix without major renovation efforts and costs. Moisture problems arise for a variety of reasons, including roof leaks, rain penetration through leaky windows, envelope design and construction defects such as low-permeability wall coverings in hot and humid climates, and poor building pressure control. These problems are largely avoidable but require an understanding of building moisture movement and attention to detail in envelope design and construction and in mechanical system selection, installation, and operation.

### *Poor Outdoor Air Quality*

As noted previously, the traditional means of dealing with IAQ is through outdoor air ventilation. While ventilation can be an effective means to dilute indoor contaminants, it assumes that the outdoor air is cleaner than the indoor air. In many locations and for many contaminants, this is not the case, and insufficiently treated ventilation air can actually make IAQ worse. Poor outdoor air quality includes regionally elevated outdoor contaminant levels as well as local sources, such as motor vehicle exhaust from nearby roadways and contaminants generated by activities in adjacent buildings. Some programs encouraging higher levels of building performance recommend increasing outdoor air ventilation rates, but such recommendations should be based on the consideration of the potential impacts of poor outdoor air quality. ASHRAE Standard 62.1 requires the assessment of outdoor air quality in the vicinity of a building and requires outdoor air cleaning under some circumstances. Given the key role of outdoor air ventilation in IAQ control, this Guide covers outdoor air quality and air cleaning alternatives in detail.

### *Moisture and Dirt in Ventilation Systems*

Dirt accumulation in ventilation systems, combined with poor management of water, can lead to biological growth in the airstream and serious IAQ problems. These conditions generally result from inadequate levels of particle filtration, poor filter maintenance, and problems with cooling coil condensate or other moisture sources. ASHRAE Standard 62.1 contains several requirements related to dirt and moisture management in ventilation systems. Given the seriousness of the problems that can result, this Guide addresses the topic in more detail.

### *Indoor Contaminant Sources*

Many IAQ problems are associated with indoor contaminant sources that are unusually strong or otherwise cannot be handled by typical or code-compliant levels of outdoor air ventilation. Many contaminants are released by normal building materials and furnishings, especially when new, and also by materials and substances brought into the building during operation. Unusual, unexpected, or atypically high contaminant emissions from indoor sources are associated with many IAQ problems, and this Guide speaks to the issues of material selection, cleaning, and other indoor sources.

### *Contaminants from Indoor Equipment and Activities*

The wide range of occupancies and activities in commercial and institutional buildings involve many different types of equipment and activities. IAQ problems have resulted from improper equipment operation, inadequate exhaust ventilation, and poor choices of materials used in some of these activities. This Guide contains information on how to decrease the likelihood of such problems.

### *Inadequate Ventilation Rates*

While building codes and standards have addressed outdoor air ventilation for decades, many buildings and spaces are poorly ventilated, which increases the likelihood of IAQ problems. There are a variety of reasons for inadequate ventilation rates, including lack of compliance with applicable codes and standards, installation or maintenance problems that lead to the design ventilation rate not being achieved in practice, or space use changes without an assessment of the need for updated ventilation rates. Also, system-level outdoor air intake rates may be adequate, but air distribution problems can lead to certain areas in the building being poorly ventilated. While ASHRAE Standard 62.1 covers the determination of design ventilation rates, additional guidance is provided in this Design Guide to help address these issues.

### *Ineffective Filtration and Air Cleaning*

Filtration and air cleaning are effective means of controlling many indoor air pollutants, particularly those associated with poor outdoor air quality. Air filtration or air cleaning, therefore, can provide an important adjunct, and in some cases substitute, for outdoor air ventilation. This Guide provides a detailed treatment of filtration and air-cleaning alternatives that, when properly administered and maintained, can improve both IAQ and energy performance.

## SCOPE: What Is and Isn't Covered in this Document?

As noted previously, this document addresses the design and construction of commercial and institutional buildings, including but not limited to office, retail, educational, lodging, and public assembly buildings, with no restrictions as to the building sizes or system types to be covered. These buildings are the same as those covered by ASHRAE Standard 62.1 and are the focus of the bulk of the recommendations in this Guide.

The scope of this Guide is necessarily limited due to both the resources available for its development and the practical need to bound the effort so that could be completed in a reasonable amount of time. Other IAQ issues and other spaces types still need to be considered, and ideally guidance will be provided for these through other efforts in the future.

Several space types and issues are not covered directly in terms of providing specific design guidance, but this Guide does attempt to address their interactions with the rest of the building and other systems. These include commercial kitchens, medical procedure rooms, natatoriums, cold buildings such as cold storage facilities and ice arenas, and laboratory, residential, and industrial spaces.

Multiple chemical sensitivity is not specifically addressed in this Guide. However, improved IAQ will benefit those who experience this condition. The National Institute of Building Sciences (NIBS) recently published a report for the U.S. Access Board that speaks directly to these concerns and contains detailed recommendations to accommodate individuals who experience these sensitivities. That report is available at http://ieq.nibs.org (NIBS 2006).

Extraordinary incidents, both natural (earthquakes, fire, floods) and intentional (terrorist attacks) are not addressed in this Guide. Information on design and planning for such events are available from a number of sources, including Federal Emergency Management Agency (FEMA, www.fema.gov) and National Fire Protection Association (NFPA, www.nfpa.org) documents.

This Guide does not address indoor smoking, as it is incompatible with good IAQ based on the health risks associated with environmental tobacco smoke and the inability of engineering controls to adequately control those risks (see the 2008 ASHRAE Position Document on Environmental Tobacco Smoke at www.ashrae.org/docLib/20090120_POS_ETS.pdf for more information and references) (ASHRAE 2008a).

## How This Guide is Organized

Based on the known causes of the IAQ problems discussed in this introduction, this Guide is organized around eight Objectives for improving building IAQ:

Objective 1 – Manage the Design and Construction Process to Achieve Good IAQ
Objective 2 – Control Moisture in Building Assemblies
Objective 3 – Limit Entry of Outdoor Contaminants
Objective 4 – Control Moisture and Contaminants Related to Mechanical Systems
Objective 5 – Limit Contaminants from Indoor Sources
Objective 6 – Capture and Exhaust Contaminants from Building Equipment and Activities
Objective 7 – Reduce Contaminant Concentrations through Ventilation, Filtration, and Air Cleaning
Objective 8 – Apply More Advanced Ventilation Approaches

Within each Objective are several Strategies designed to help achieve that Objective.

### *How to Use this Guide*

Starting with the eight Objectives and the Strategies for each, the information in this Guide is broken into summary guidance (Part I) and detailed guidance (Part II). Both Part I and Part II are included in the electronic version of this Guide; only Part I is included in the printed version of this Guide.

### *Part I—Summary Guidance*

**Objectives and Strategies.** An overview for each Objective in Part I provides an understanding of why the Objective is important for good IAQ. Each overview is followed by brief descriptions of the Strategies that can be employed to achieve that Objective. An objective graphic for each Objective provides a visual reference to the Strategies intended to achieve the Objective. In the electronic version of this Guide, the objective graphic contains blue interactive links to the summary guidance for each Strategy in Part I.

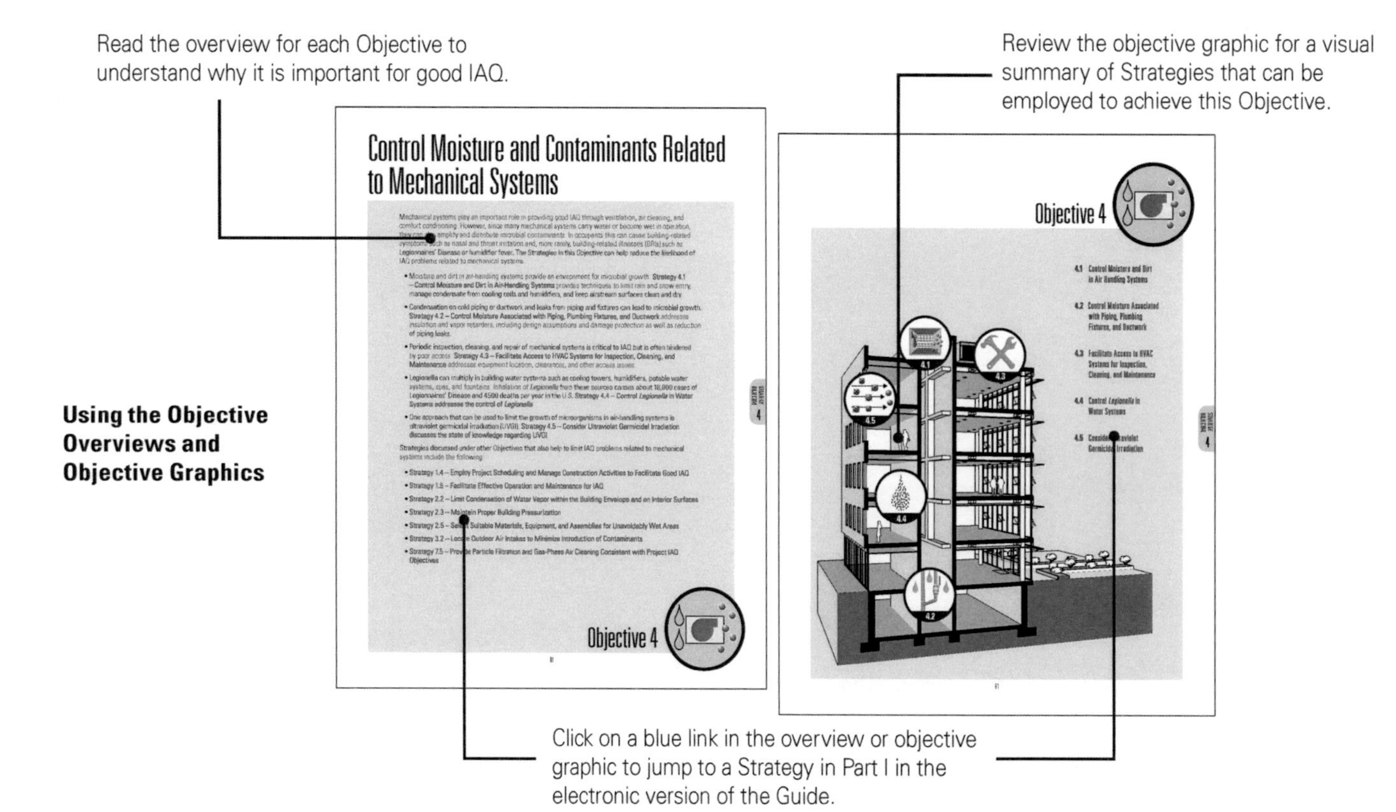

Each Strategy in Part I has an overview that describes why the Strategy is important, how to determine whether it needs to be considered in a particular project, and the general nature of the solutions. Each Strategy contains tabular and graphical guides to the detailed information in Part II that act as roadmaps and outline specific elements to be considered when implementing each Strategy. The tabular guides can also be used as checklists in project planning. The graphical guide provides a visual reference for each element of the Strategy. In the electronic version of this Guide, both the tabular and graphical guides in Part I contain blue interactive links that take the reader directly to the corresponding detailed information for each Strategy in Part II.

All sources cited in the Objectives and Strategies in Part I are listed in the single References section at the end of Part I.

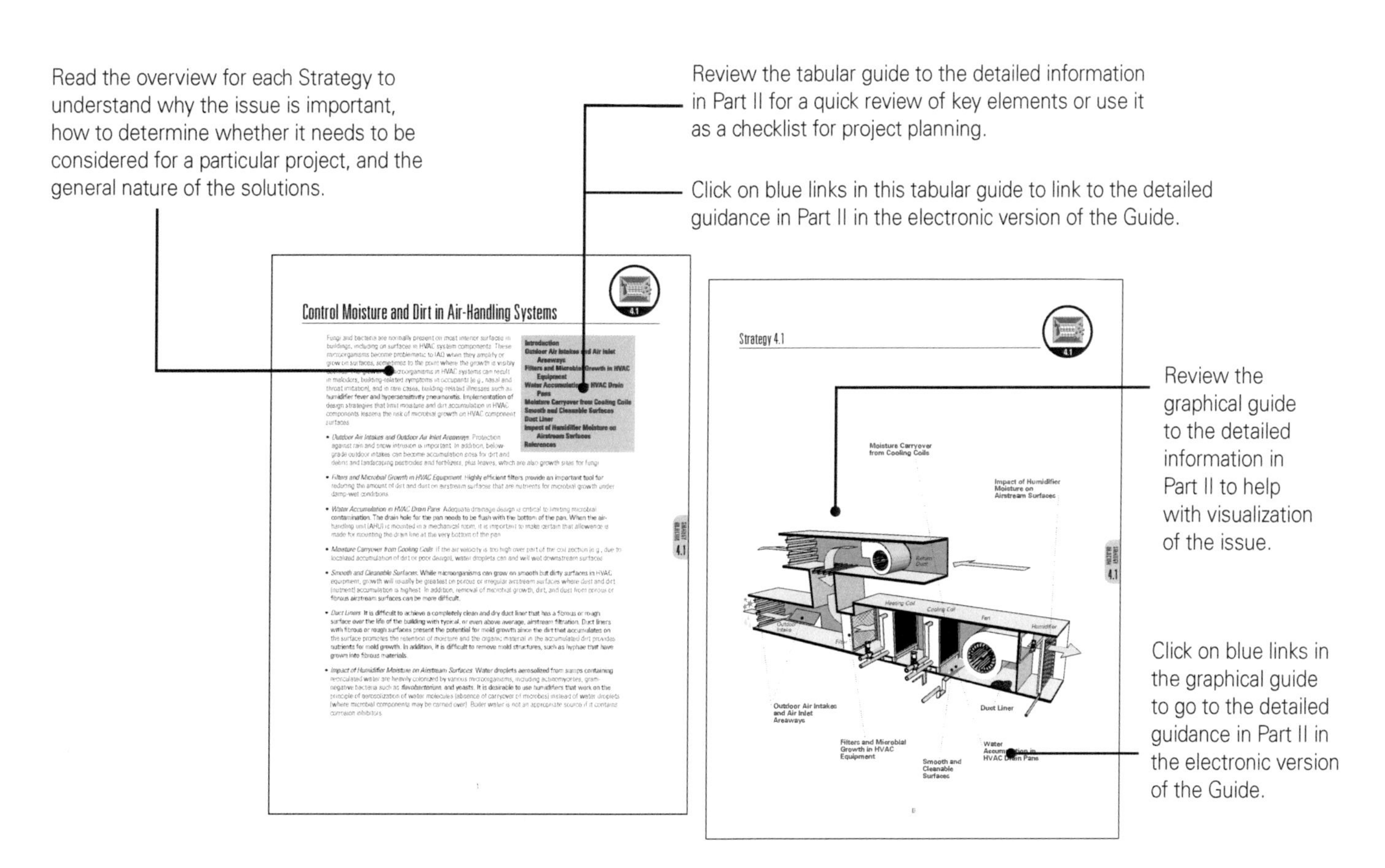


**Using the Strategy Overviews and Tabular and Graphical Guides**

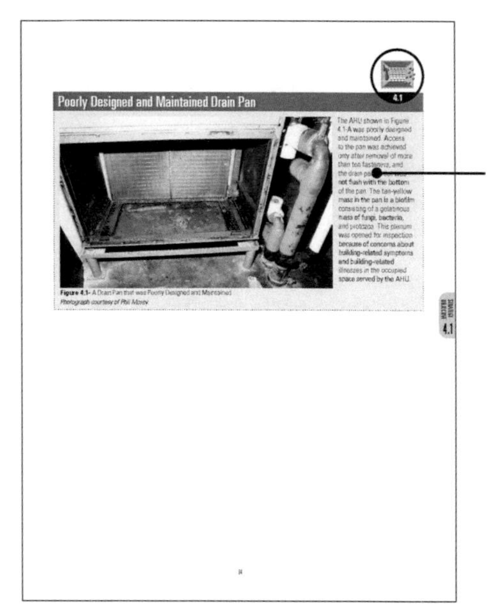


Check out the case studies for insights from other projects.

**Using the Case Studies**

**Case Studies.** Case studies are included in the summary discussions of the Strategies in Part I. They provide insights into the IAQ problems and successes others have had in dealing with each topic.

**Sidebars.** There are sidebars interspersed throughout Part I to provide further explanation of code guidance or of terms or examples.

***Part II—Detailed Guidance***

Whether you've already had projects with moisture problems and want guidance on how to avoid them, are doing your first brownfield project with the potential for vapor intrusion, have been asked to design a green building with low-emitting materials, want to do a better job of controlling outdoor pollutants in an urban location, or want to reduce the energy costs due to ventilation, the Part II detailed guidance will help.

The Part II detailed guidance provides information to use when working on a specific project, including detailed design, construction, and Cx recommendations; calculation procedures; additional case studies and sidebars; extensive references; and more. Part II is structured by Objectives and Strategies identical to the Part I summary guidance. Each Strategy in Part II concludes with its own references section so that the source information is close at hand when dealing with a particular issue.

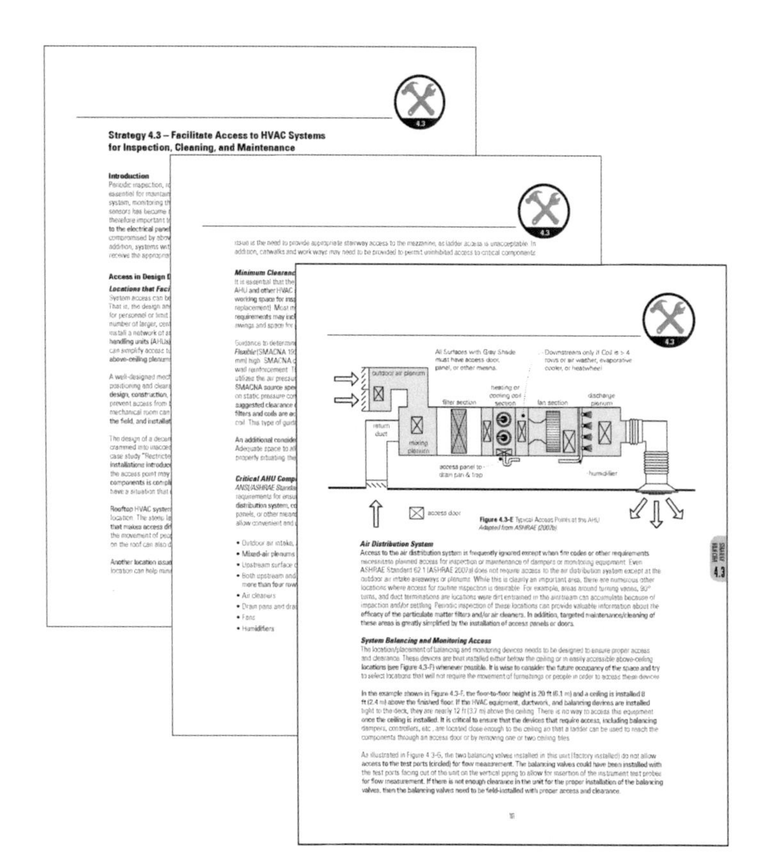


**Using the Part II Detailed Guidance**

# PART I—Summary Guidance

## Overview Information for Design, Construction, and Commissioning for IAQ

Part I of this Guide provides a convenient summary of the key elements of design for indoor air quality (IAQ). These are grouped into eight Objectives:

- Objective 1 – Manage the Design and Construction Process to Achieve Good IAQ
- Objective 2 – Control Moisture in Building Assemblies
- Objective 3 – Limit Entry of Outdoor Contaminants
- Objective 4 – Control Moisture and Contaminants Related to Mechanical Systems
- Objective 5 – Limit Contaminants from Indoor Sources
- Objective 6 – Capture and Exhaust Contaminants from Building Equipment and Activities
- Objective 7 – Reduce Contaminant Concentrations through Ventilation, Filtration, and Air Cleaning
- Objective 8 – Apply More Advanced Ventilation Approaches

Overview text provides an introduction to each Objective. The overviews are followed by descriptions of the Strategies that can be employed to achieve each Objective. An objective graphic for each Objective identifies major Strategies related to that Objective. In the electronic version of this Guide, each objective graphic in Part I contains blue interactive links to the Strategies for that Objective in Part I.

For each Strategy, there is an overview, both tabular and graphical guides to the detailed information in Part II, one or more case studies, and occasionally a sidebar. The overview explains why the issue is important for IAQ, how to determine whether it needs to be considered for a particular project, and potential design solutions to the problem. In Part I, each Strategy's tabular guide to the detailed information in Part II identifies the elements that need to be addressed in meeting each Objective and can be modified to be used as a checklist in project planning. Each Strategy's graphical guide to the detailed information in Part II provides a visual overview of the issue. In the electronic version of this Guide, both the tabular and graphical guides in Part I contain blue interactive links to the Part II detailed guidance.

# Manage the Design and Construction Process to Achieve Good IAQ

The single most important step an owner or design team leader can take to reliably deliver good IAQ is to use effective project processes. Lacking these, even the most sophisticated suite of IAQ technologies may not deliver the desired results. Using effective project processes, however, even simple designs can avoid IAQ problems and provide a good indoor environment.

- Strategy 1.1 – Integrate Design Approach and Solutions describes approaches to integrate design across disciplines, enabling achievement of IAQ and other performance goals at lower cost. Many IAQ problems occur because building elements are designed by different disciplines working in relative isolation. Even design elements that do not appear to be related can sometimes interact in ways that are detrimental to IAQ.
- Strategy 1.2 – Commission to Ensure that the Owner's IAQ Requirements are Met provides guidance on commissioning (Cx) as a quality control process for IAQ, from establishment of the owner's IAQ requirements at project inception to construction observation and functional testing. For buildings as for any other product, quality control in design and execution is necessary to achieve the desired result.
- Strategy 1.3 – Select HVAC Systems to Improve IAQ and Reduce the Energy Impacts of Ventilation explains how the type of HVAC system selected can constrain the level of IAQ achievable by limiting the capability for filtration, space humidity control, building pressurization, or separation of intakes from contaminant sources. It can also have a major impact on the energy required for ventilation. Yet the type of system is often selected by the architect before the engineer is involved or chosen based on cost, space required, or other factors without adequate consideration of IAQ implications.
- Strategy 1.4 – Employ Project Scheduling and Manage Construction Activities to Facilitate Good IAQ highlights the importance of construction processes to IAQ. A project schedule that is too compressed or improperly sequenced can jeopardize IAQ. Likewise, failure to manage contaminants and water during construction can have a detrimental effect on occupants in buildings undergoing renovation and on long-term IAQ in new buildings.
- Strategy 1.5 – Facilitate Effective Operation and Maintenance for IAQ explains how the design team can help O&M staff deliver performance consistent with the design intent through appropriate system selection, system-oriented documentation, and system-oriented training.

# Objective 1

STRATEGY
OBJECTIVE
1

**1.1** Integrate Design Approach and Solutions

**1.2** Commission to Ensure that the Owner's IAQ Requirements are Met

**1.3** Select HVAC Systems to Improve IAQ and Reduce the Energy Impacts of Ventilation

**1.4** Employ Project Scheduling and Manage Construction Activities to Facilitate Good IAQ

**1.5** Facilitate Effective Operation and Maintenance for IAQ

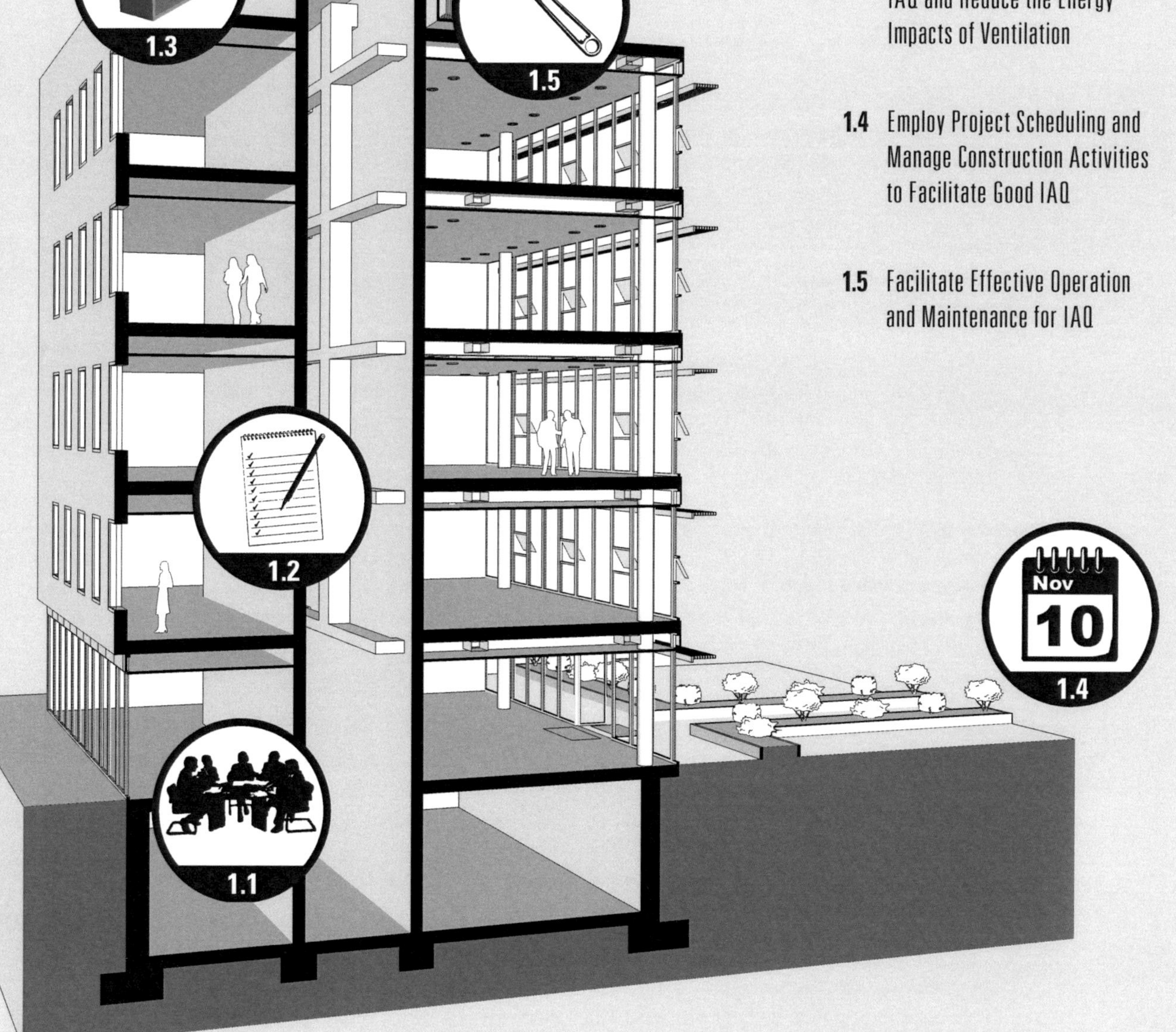

# Integrate Design Approach and Solutions



Integrated design is one of today's buzzwords in "green" and "sustainable" building design, but nowhere is it more important or valuable than in relation to IAQ. Most if not all of the design approaches and solutions that are important for achieving good IAQ are also important for thermal comfort and energy efficiency. They also have very strong connections to illumination and acoustics.

Thermal comfort and good IAQ are intricately bound together both in the characteristics of the indoor environment and in the way building occupants respond to the indoor environment. Occupants' perceptions of the indoor environment and the indoor environmental quality (IEQ) impacts on occupant health have strong interactions and, in reality, cannot be separated. Beyond environmental control systems, IAQ is strongly determined by the building structure and envelope, so all key members of the design team play a role in determining the potential for achieving good IAQ in your designs.

The team members responsible for the ventilation and thermal control solutions affect and are affected by the acoustic and illumination requirements and solutions. Noise from mechanical systems, waste heat from electrical illumination sources, or heat loss or gain through glazing are as important to the selection of ventilation solutions as the pollutant loads coming from building materials, occupant activities, building equipment, appliances, or any other sources. Only by considering all of the potential loads can the optimal solution for ventilation, material selection, and envelope design be made effectively.

The easiest and most effective way to accomplish integrated design is to assemble the entire design team at the beginning of the project and to brainstorm siting, overall building configuration, ventilation, thermal control, and illumination concepts as a group. The give and take of the initial design charette with the key members present will help each team member to appreciate the specialized concerns of the others and enable the group to develop a solution that best integrates everyone's best ideas.

Once the initial design concept is agreed upon, then the evolution of the design through its various stages can occur with a shared concept and the potential for direct interaction among team members as challenges arise later in the process. The design team leader ultimately must make decisions when conflicts arise, but starting with a concept shared by the whole team will minimize the number and importance of those conflicts later in the process.

In typical design processes, lacking such a collaborative effort to produce an overall design concept, the building's basic concept ends up reflecting only some of the important considerations. Then the remainder of the design process looks more like an effort to retrofit the design concept to accommodate the concerns ignored initially. It is also true that many of the most environmentally responsible design solutions can work together to produce a synergy that is not achieved when such collaboration and integration is absent. Reducing loads—whether of pollutant emissions or of heat gain or loss—reduces demand for ventilation and conditioning of outdoor air and results in lower first costs for equipment as well as lower operating costs.

Building design professionals understand that the design of virtually every building element affects the performance of other elements, so it makes sense to integrate various design elements of a building. Unfortunately, the prevailing design process of our time tends to create design elements in a compartmentalized and linear process rather than jointly designing these elements in an interactive process. Figure 1.1-A depicts the traditional design team; Figure 1.1-B depicts the integrated design team.

# Strategy 1.1

**Current Trends call for Integrated Design**

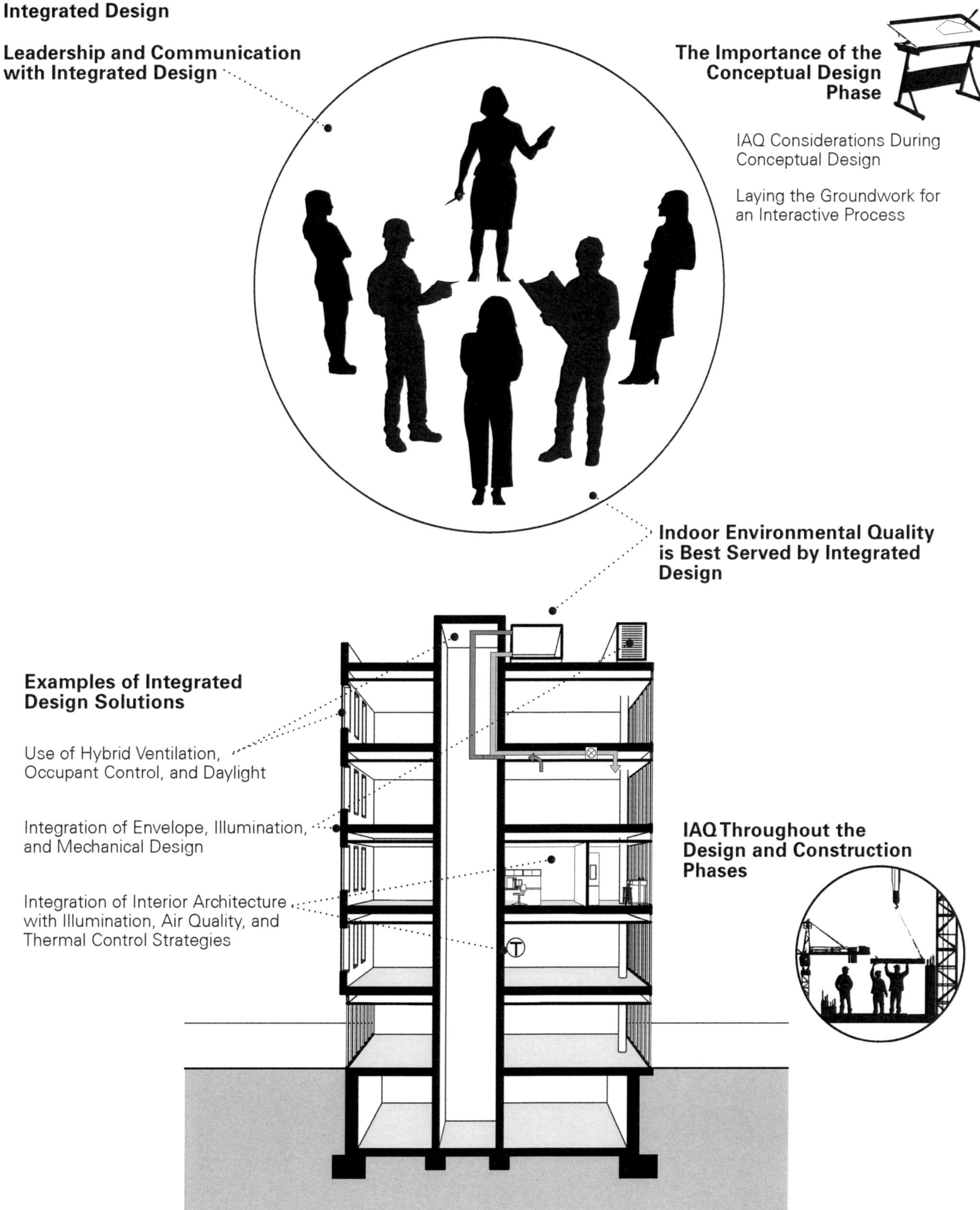

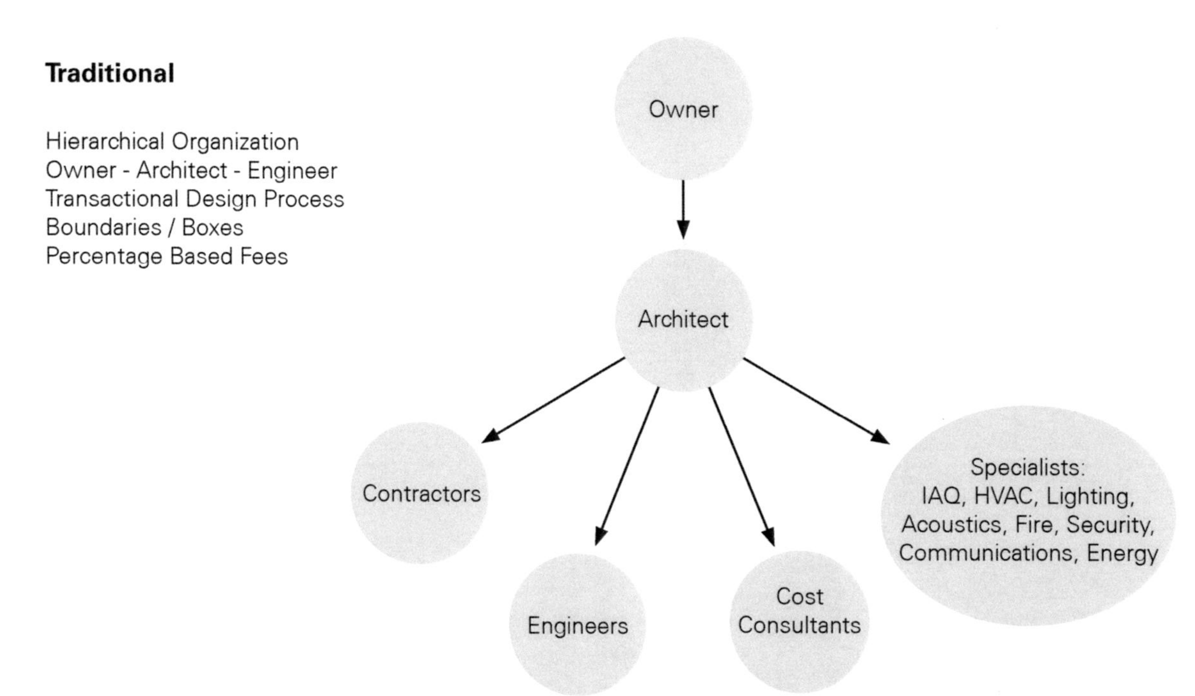


**Figure 1.1-A** Traditional Design Team

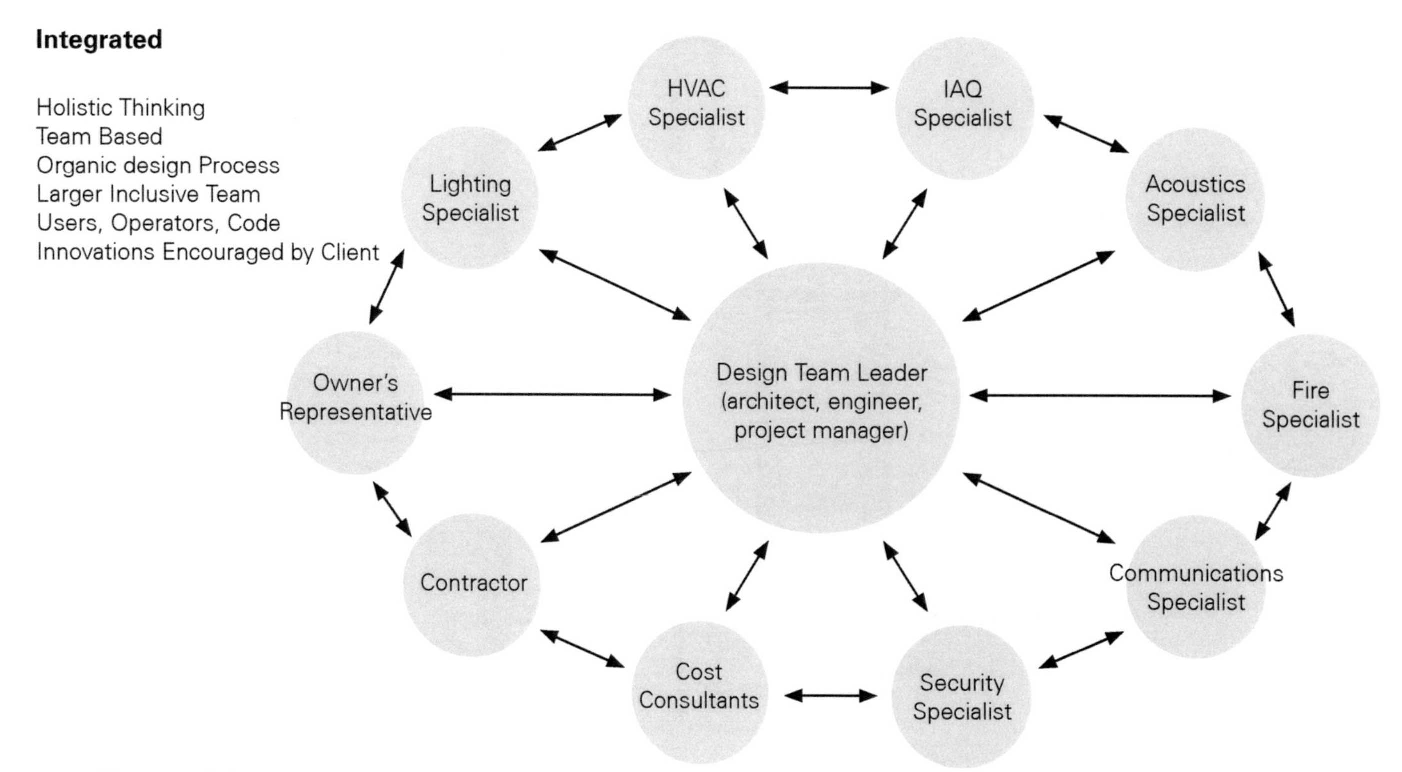


**Figure 1.1-B** Integrated Design Team

## Integrated Design Process



Typically, an integrated design process begins with a "charette"—a gathering of the major players, often including the client or future occupants. A design charette may also be a gathering of the key members of the design team including all major consultants. When the focus is on environmental performance, indoor environment, energy, and environmental impacts, it is common to identify the major issues and establish goals very early in the process, ideally during the generation of the conceptual or schematic design.

An example of this process is much of the work of architect Bob Berkebile, FAIA, principal of the firm Berkebile Nelson Immenschuh McDowell Architects in Kansas City, Missouri. Berkebile founded the American Institute of Architects (AIA) Committee on the Environment and has long been one of the leading practitioners of environmentally responsible design including IEQ, energy performance, and general environmental impacts. Figures 1.1-B and 1.1-C show typical gatherings in the integrated design practice of BNIM.

**Figure 1.1-B** Designers Choosing Materials at a Typical Charette
*Photograph copyright BNIM.*

**Figure 1.1-C** Design Team Studying a Model of an Early Design Concept at a Charette
*Photograph copyright BNIM.*

# Commission to Ensure that the Owner's IAQ Requirements are Met

**Introduction**
**Pre-Design Phase Commissioning**
- Commissioning Team: Specialists Needed for IAQ Items
- Owner's Project Requirements for IAQ
- Commissioning Scope and Budget Related to IAQ
- Special Project Schedule Needs for IAQ

**Design Phase Commissioning**
- IAQ Basis of Design (BoD)
- Design Review for IAQ
- Construction Process Requirements
- Construction Checklists for IAQ

**Construction Phase Commissioning**
- Coordination for IAQ
- Review of Submittals for IAQ
- Construction Observation/Verification for IAQ
- Functional Testing for IAQ
- Systems Manual and O&M Training for IAQ

**Occupancy and Operations**
**References**

**What is Commissioning and Why Is It Needed?**
Few manufacturers today would consider producing a product without a formal quality control process. Yet the majority of buildings are built without the use of systematic quality control procedures. As a result, buildings may be turned over with undetected deficiencies, and key assemblies or systems may fail to function as intended. IAQ may suffer due to any number of problems in design, material, and equipment selection or construction. To address these problems, a growing number of building owners are incorporating commissioning (Cx), a quality-focused process that is used to complete successful construction projects (ASHRAE 2005).

**Commissioning Starts at Project Inception**
It is a common misconception that Cx is a post-construction process. In fact, Cx needs to start in the pre-design phase to maximize its effectiveness and cost-effectiveness. During this phase, the owner should select a commissioning authority (CxA) and establish the Cx scope and budget. The design team's responsibilities related to Cx need to be defined in their agreements with the owner.

During pre-design, the CA helps the owner identify and make explicit all functional requirements for the project. These requirements then become the focus of the Cx process. For example, every owner expects his or her building to be free of condensation and mold problems, be properly ventilated, and provide good-quality ventilation air, but the team can lose focus on these Owner's Project Requirements (OPR) so that they fail to be met if the OPR are not explicitly stated and tracked throughout the project.

The CA needs to provide input to the project schedule to ensure that it accommodates the steps necessary to achieve the owner's IAQ requirements. This input may include, for example, the timing of inspections that must be made while key assemblies are still open or the proper sequencing of work to avoid moisture damage.

**Commissioning the Design**
It is much easier and cheaper to correct deficiencies on paper during design than in the finished building after construction.

During conceptual design, Cx calls upon the design team to record the concepts, calculations, decisions, and product selections used to meet the OPR and applicable codes and standards in a Basis of Design (BoD) document. The CA plays an important role in reviewing this BoD document to determine whether it will meet the owner's requirements. The CA continues to review the design in the design development and construction documents phases to ensure that they will fulfill the owner's needs.

The CA assists the design team in incorporating into the specifications the Cx work that will be required of contractors so that the contractors can understand and budget their role in the Cx process.

**Commissioning the Construction**
During construction, the CA monitors work to ensure that it does not compromise the OPR. This includes reviewing submittals to ensure that they are consistent with the OPR and BoD and that they provide for Cx needs. It also includes early and ongoing observation of key aspects of construction to ensure that the owner's requirements are not compromised. This may include, for example, checking the continuity of drainage planes and air barriers while walls are under construction or checking that maintenance access is preserved as HVAC equipment and later services are installed.

1.2



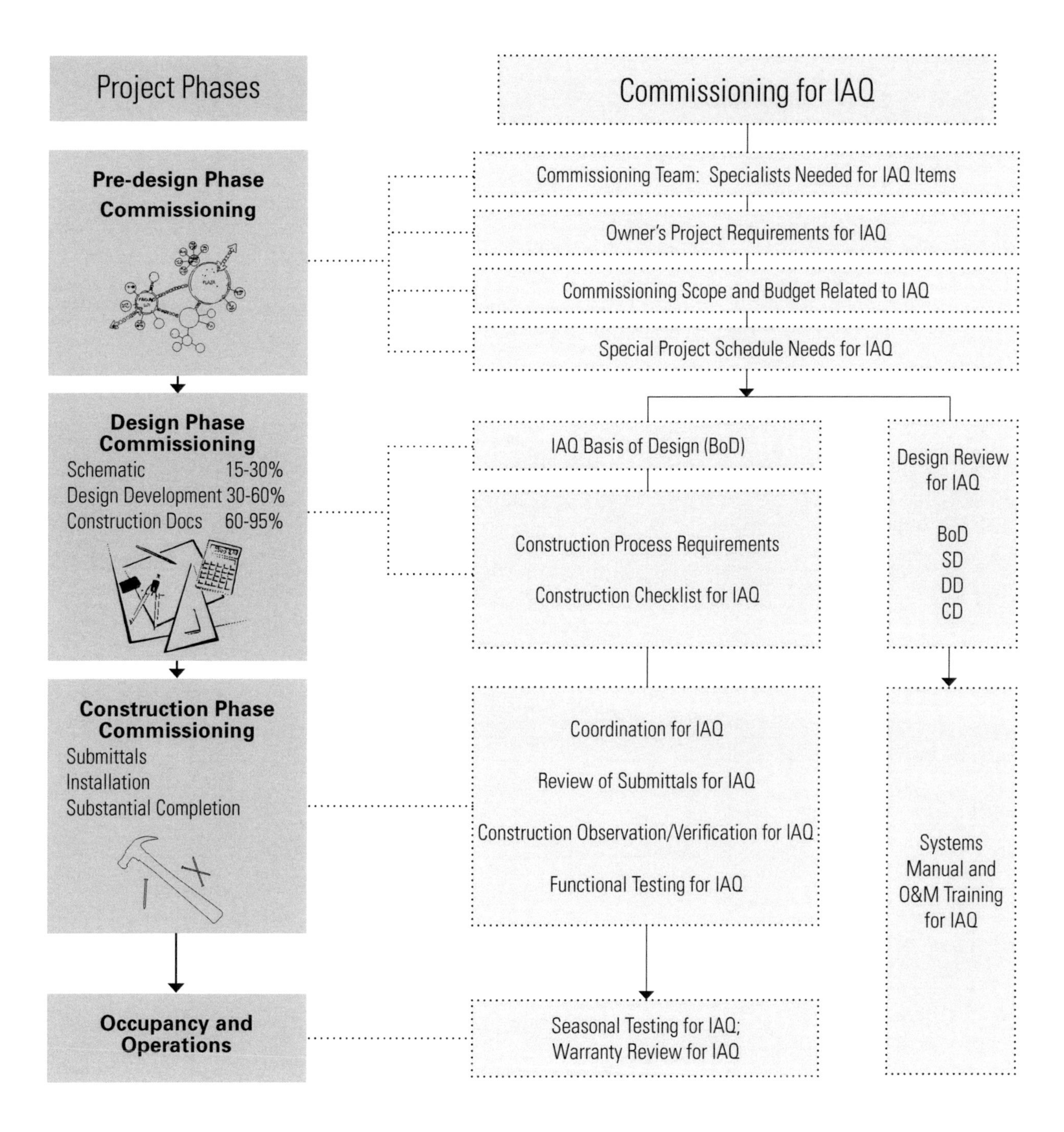
Project Phases
Commissioning for IAQ
Pre-design Phase Commissioning
Commissioning Team: Specialists Needed for IAQ Items
Owner's Project Requirements for IAQ
Commissioning Scope and Budget Related to IAQ
Special Project Schedule Needs for IAQ
Design Phase Commissioning
Schematic 15-30%
Design Development 30-60%
Construction Docs 60-95%
IAQ Basis of Design (BoD)
Construction Process Requirements
Construction Checklist for IAQ
Design Review for IAQ
BoD
SD
DD
CD
Construction Phase Commissioning
Submittals
Installation
Substantial Completion
Coordination for IAQ
Review of Submittals for IAQ
Construction Observation/Verification for IAQ
Functional Testing for IAQ
Systems Manual and O&M Training for IAQ
Occupancy and Operations
Seasonal Testing for IAQ;
Warranty Review for IAQ

The CA also provides the installation and start-up checklists that are executed and signed by the contractors and spot-checks them after completion. Often the CA verifies a statistical sample of the balancing report by observing the balancer as he or she conducts repeated measurements.

### Testing for Acceptance

The CA designs, oversees, and documents functional tests that determine the ability of building assemblies and systems to meet the OPR. These may include testing of building assemblies for water penetration and air leakage or testing of control system sequences of operation for proper performance.

### Systems Manual and O&M Training

O&M manuals are often massive and can lack key information while being laden with material that does not apply to the project. O&M training is often a cursory afterthought. In commissioned projects, however, the CA often defines requirements for a systems manual and O&M training to support ongoing achievement of the OPR and verifies their delivery to O&M staff.

## Commissioning to Ensure that Design Ventilation Rates Are Met

Poor ventilation in an extensively renovated theater led to patron complaints of stuffiness. Investigation identified several factors contributing to low ventilation rates:

- The design called for demand-controlled ventilation (DCV) based on carbon dioxide ($CO_2$), with minimum outdoor air (OA) flow modulated between 2000 and 7200 cfm (940 and 3400 L/s) to maintain $CO_2$ at or below

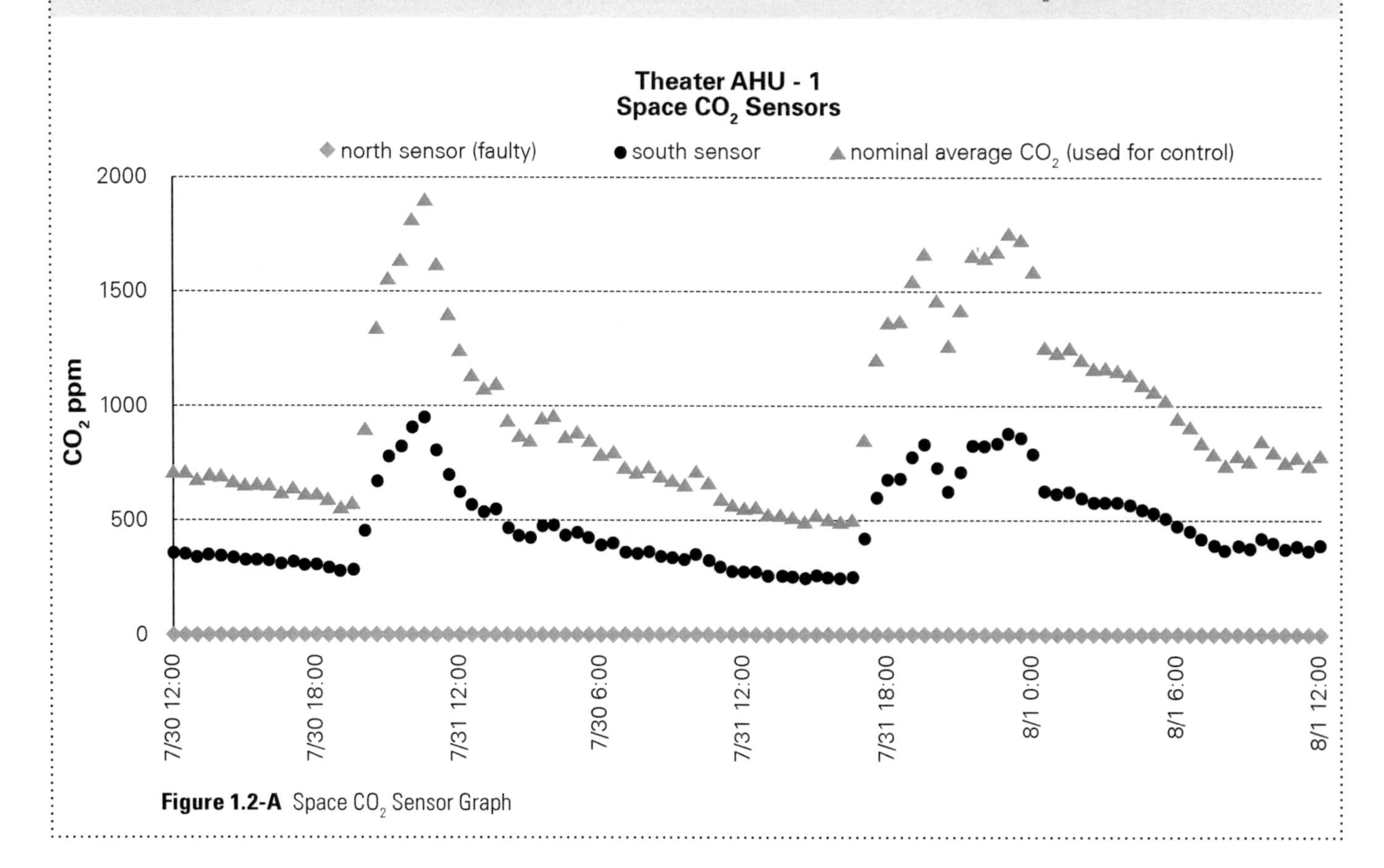


**Figure 1.2-A** Space $CO_2$ Sensor Graph

900 ppm (1600 mg/m³). The programming as implemented actually reset the minimum OA between 2000 cfm (940 L/s) at 1000 ppm (1800 mg/m³) $CO_2$ and 3000 cfm (1400 L/s) at 1500 ppm (2700 mg/m³) $CO_2$.

- The north $CO_2$ sensor (diamonds near zero in Figure 1.2-A) was faulty and consistently read about 4 ppm (7 mg/m³). The average $CO_2$ concentration used by the control system to adjust OA flow thus appeared to be about half its actual value. For example, if the actual $CO_2$ was 1900 ppm (3400 mg/m³), the average value passed to the control loop was (1900 + 4)/2 = 952 ppm [(3400 + 7)/2 = 1703 mg/m³], and the reset only called for the system to deliver 2000 cfm (940 L/s) of OA.
- The OA measuring station read about 1100 cfm (520 L/s) with the fan off, so it reached a reading of 2000 cfm (940 L/s) at an actual flow below 2000 cfm (940 L/s).

As a result of these problems, the actual minimum OA never went above about 2000 cfm (940 L/s) (Figure 1.2-B). Because the project was not commissioned, the cause of the stuffiness was not diagnosed until several years after the renovation when the building was retro-commissioned.

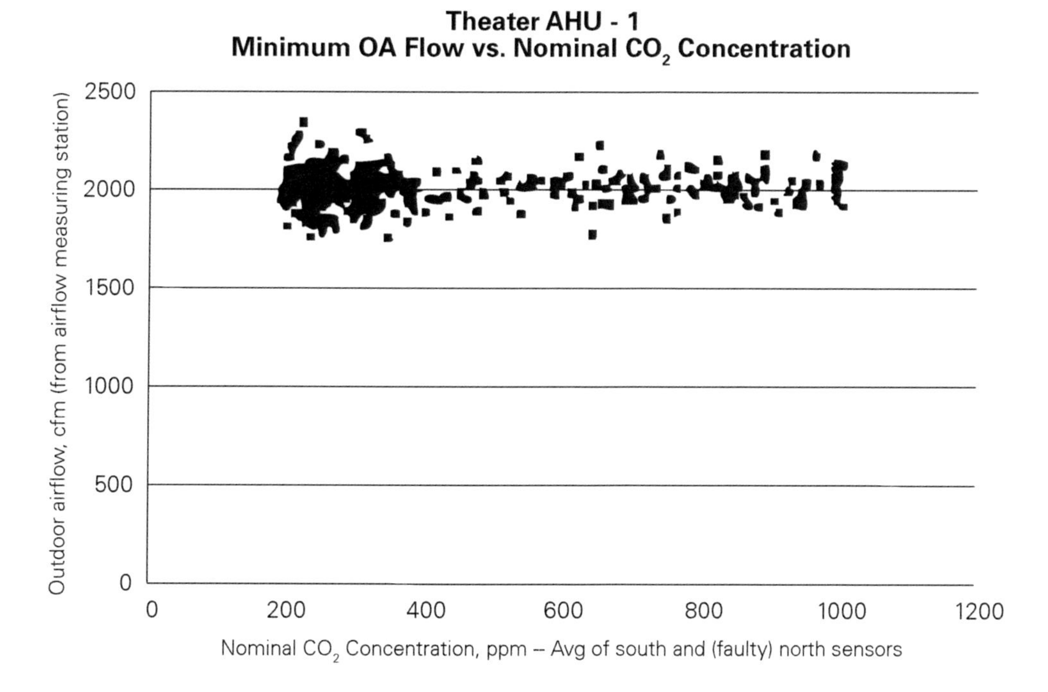


**Figure 1.2-B** Minimum OA Flow vs Nominal $CO_2$ Concentration

STRATEGY OBJECTIVE 1.2

# Select HVAC Systems to Improve IAQ and Reduce the Energy Impacts of Ventilation

Prior to the invention of electrical power and mechanical cooling, buildings were designed to take full advantage of the prevailing natural forces such as wind, outdoor temperature, sunlight, and thermal mass to ventilate and cool occupied spaces. After the invention of mechanical air conditioning, buildings could be designed with little concern for the building configuration, solar gains, or outdoor conditions. Today, due to concerns about global warming, there is a worldwide movement toward buildings that employ sustainable strategies. Building owners, architects, and engineers are designing buildings that are more sympathetic with their surrounding environment and using natural features and forces to reduce a building's "environmental footprint." This emerging trend is forcing designers to look at energy in a different way, giving consideration to the quantity of energy used in buildings and also the quality of the energy used. An example of this type of design is using passive/natural or low-grade energy/waste/renewable energy in lieu of fossil fuel/electrical power (Willmert 2001).



The primary difference between present-day conventional design and this emerging approach is in the design process. Designing a building with a reduced environmental footprint requires a fully integrated design process in which all members of the design team work in an integrated framework, thinking about all design decisions within the context of occupant and building safety, thermal comfort, IAQ, and the impact of the design decision upon the environment. A key example of this is the use of displacement ventilation. If displacement ventilation is going to be used effectively in a building, the design team must factor solar gains and thermal envelope loads into the discussion about the ventilation strategy.

Despite these trends, HVAC system designers are still often required, for numerous reasons, to use conventional HVAC systems such as variable-air-volume (VAV) with reheat, constant volume (CV) with reheat, fan-coil (FC), or packaged HVAC equipment to condition institutional, commercial, and residential building air. Clearly these conventional systems have their place, and those designers using them can apply low-energy and improved IAQ principles within their design, effectively moving away from inefficient non-integrated design practice.

It is well understood that in mechanically ventilated buildings, HVAC systems can have a significant effect on IAQ, energy use, and occupants' well-being. Studies have shown that poor IAQ can be directly linked to insufficient ventilation rates and inappropriate HVAC system design or operation (Seppanen and Fisk 2004). Also, Mendell and Heath (2005) found evidence that certain conditions commonly found in U.S. schools, such as low ventilation rates, have adverse effects on the health and the academic performance of many of the more than 50 million U.S. schoolchildren.

In addition to the measurable IEQ factors such as temperature, humidity, $CO_2$, and air speed, the HVAC system design should also consider human factors such as personal control over the environment. It has been suggested in a number of studies (e.g., Wyon [1996]) that ventilated buildings with enhanced occupant control of the indoor environment have lower reported rates of sick building syndrome symptoms.

# Strategy 1.3

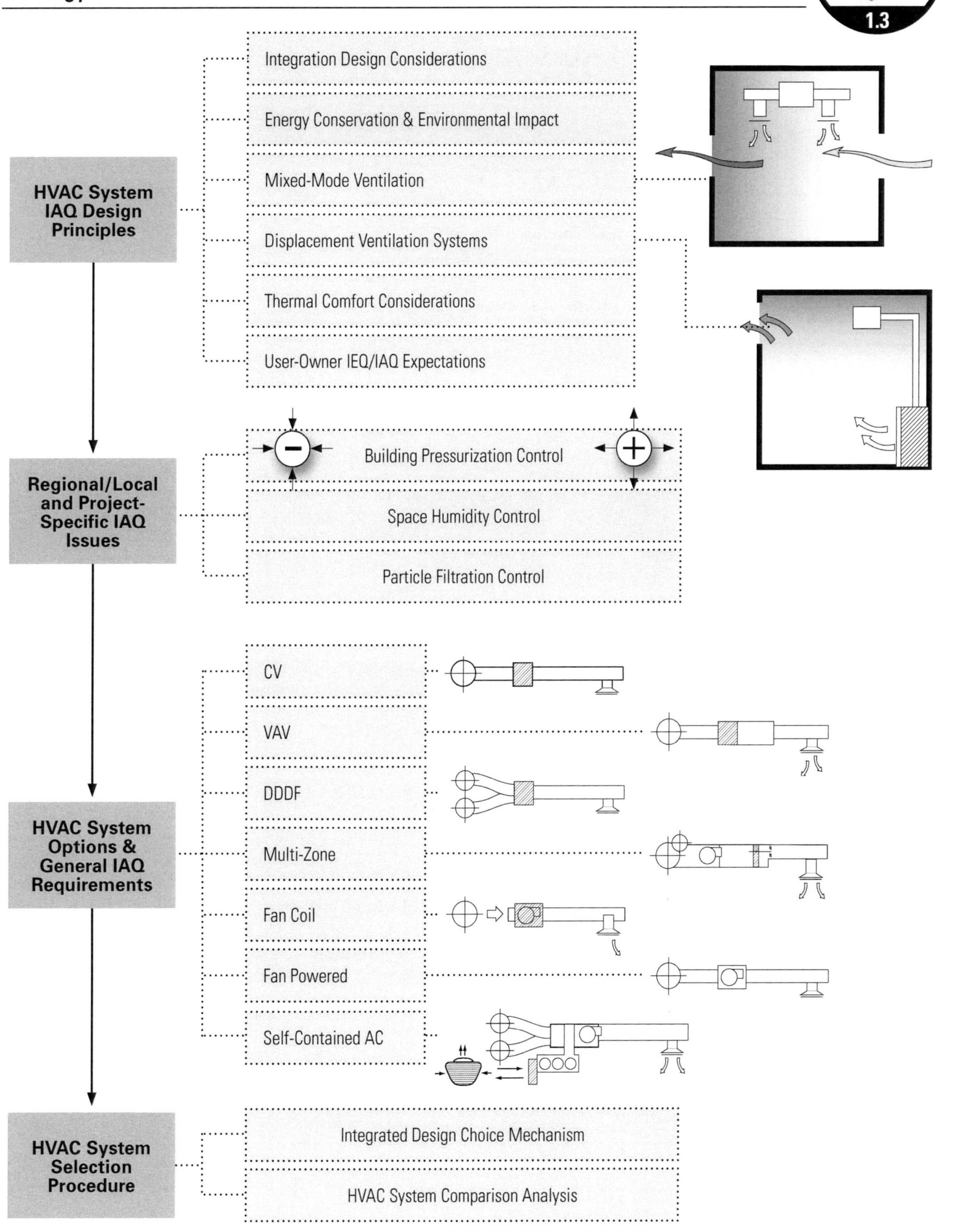

1.3
STRATEGY OBJECTIVE
1.3
HVAC System IAQ Design Principles
Integration Design Considerations
Energy Conservation & Environmental Impact
Mixed-Mode Ventilation
Displacement Ventilation Systems
Thermal Comfort Considerations
User-Owner IEQ/IAQ Expectations
Regional/Local and Project-Specific IAQ Issues
Building Pressurization Control
Space Humidity Control
Particle Filtration Control
HVAC System Options & General IAQ Requirements
CV
VAV
DDDF
Multi-Zone
Fan Coil
Fan Powered
Self-Contained AC
HVAC System Selection Procedure
Integrated Design Choice Mechanism
HVAC System Comparison Analysis

## Gulf Islands National Park Reserve Operations Centre —Sidney, British Columbia, Canada

**Figure 1.3-A** GINPR Operations Centre

*Photograph copyright Stantec Consulting Ltd. Architect McFarland Maceau Architects Ltd.*

**Figure 1.3-B** Interior View, GINPR Operations Centre

*Photograph copyright Stantec Consulting Ltd. Architect McFarland Maceau Architects Ltd.*

The Gulf Islands National Park Reserve (GINPR) Operations Centre (Figures 1.3-A and 1.3-B), designed by McFarland Marceau Architects Ltd, project architects, and Stantec Consulting Ltd., project mechanical engineers, is 11300 ft$^2$ (1050 m$^2$) and the first project in Canada to achieve the Leadership in Energy and Environmental Design (LEED) Platinum certification. The redevelopment is designed to accommodate Parks Canada and other associated user group vessels. The GINPR operations building consists of office space, a small laboratory, a library, lockers, and two sleeping quarters. The project utilizes a mixed-mode ventilation system for energy efficiency and IAQ. An ocean-based heat pump system provides heating and domestic hot water, and a rainwater collection system is used for marine wash-water and sewage conveyance. For more information on the project's energy performance and sustainability features, visit www.pc.gc.ca/pn-np/bc/gulf/ne/ne5_e.asp.

## Philip Merrill Environmental Center—Annapolis, Maryland

**Figure 1.3-C** Philip Merrill Environmental Center

*Photograph courtesy of NREL/DOE. Photographer Rob Williamson.*

As the first Leadership in Energy and Environmental Design (LEED) Platinum building in the United States, the Philip Merrill Environmental Center (Figure 1.3-C) is at the leading edge of sustainability practices. The building was designed by the architect Smith Group and was one of the 2007 winners of the Livable Buildings Awards from the Center for the Built Environment (CBE 2007).

The building, which is 32,000 ft$^2$ (2973 m$^2$), is described as a combination of space-age technology and age-old techniques. It has a large number of enhanced IEQ features, including high levels of occupant control, abundant natural lighting, and low-VOC-emission material usage.

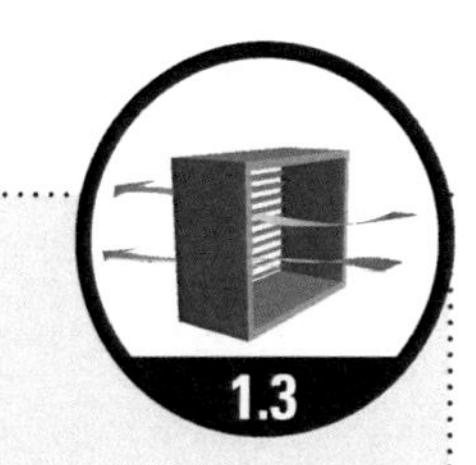


The occupants of the Philip Merrill Environmental Center were surveyed and interviewed about the IEQ and its effect on psychosocial and productivity-related factors. Key positive findings of this survey showed that occupants were highly satisfied with the building as a whole and that intangibles such as pride in the building and aesthetics contributed to high levels of morale, well-being, and a sense of belonging at work. Acoustical conditions were the most negatively rated, primarily due to distractions from people talking and loss of speech privacy associated with the highly open environment.

Figure 1.3-D shows an overall summary of survey responses regarding the Philip Merrill Environmental Center. The mean score on all of these categories was at or above 2.0 on a seven-point scale ranging from –3 to +3.

For more information on the building details of the Philip Merrill Environmental Center, visit www.archiplanet.org/wiki/Philip_Merrill_Environmental_Center.

*Source: Heerwagen and Zagrreus (2005).*



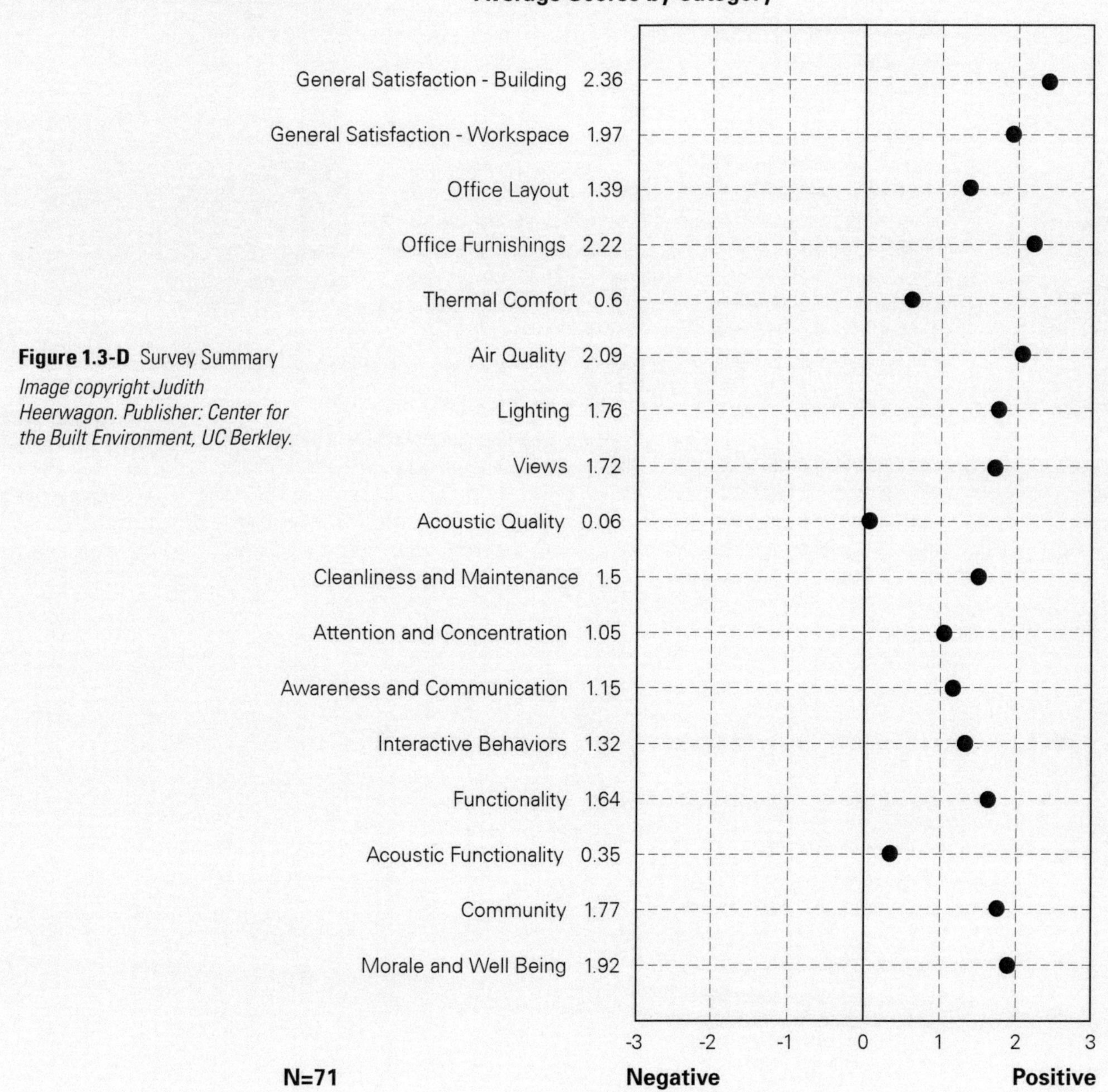


**Figure 1.3-D** Survey Summary
*Image copyright Judith Heerwagon. Publisher: Center for the Built Environment, UC Berkley.*

# Employ Project Scheduling and Manage Construction Activities to Facilitate Good IAQ

Establishing a comprehensive and realistic project construction schedule and using sound management of construction activities will help ensure the achievement of good IAQ for a building project. A building owner can have the best intentions, the design team can provide a great building design, and the contractors can achieve their best execution, but if the schedule is too compressed or has sequencing issues or if construction activities are not well managed, the IAQ of the finished project will be compromised.



All phases of the construction project need to be identified and evaluated for scheduling purposes and construction management. In particular, a Cx schedule covering each phase of the project needs to be developed, and part of the Cx work needs to include checking to insure that IAQ-related project scheduling activities are properly implemented.

Scheduling issues can be critical to achieving good IAQ. For example, installing construction materials that can absorb moisture before the building is closed in, storing construction materials in or exposing construction materials to the weather prior to installation, and starting and operating HVAC equipment for temporary heating or cooling during construction are just a few project-schedule-related items that could jeopardize the IAQ of the final building project. A sound IAQ plan coupled with proper scheduling and sequencing of the construction project will help ensure achievement of good IAQ.

The schedule needs to allow for adequate time to complete the construction activities and properly sequence them. It is best to involve the design and construction team early in the scheduling process. This could include gathering input from the construction team during the design phase to help with sequencing and planning the duration of construction activities. The schedule needs to be reviewed on a regular basis throughout the project and be maintained and updated as necessary. In addition, proper construction material selection could be evaluated to match the sequencing and scheduling of the construction activities. An accelerated project schedule can seriously undermine the achievement of good project IAQ.

STRATEGY OBJECTIVE 1.4

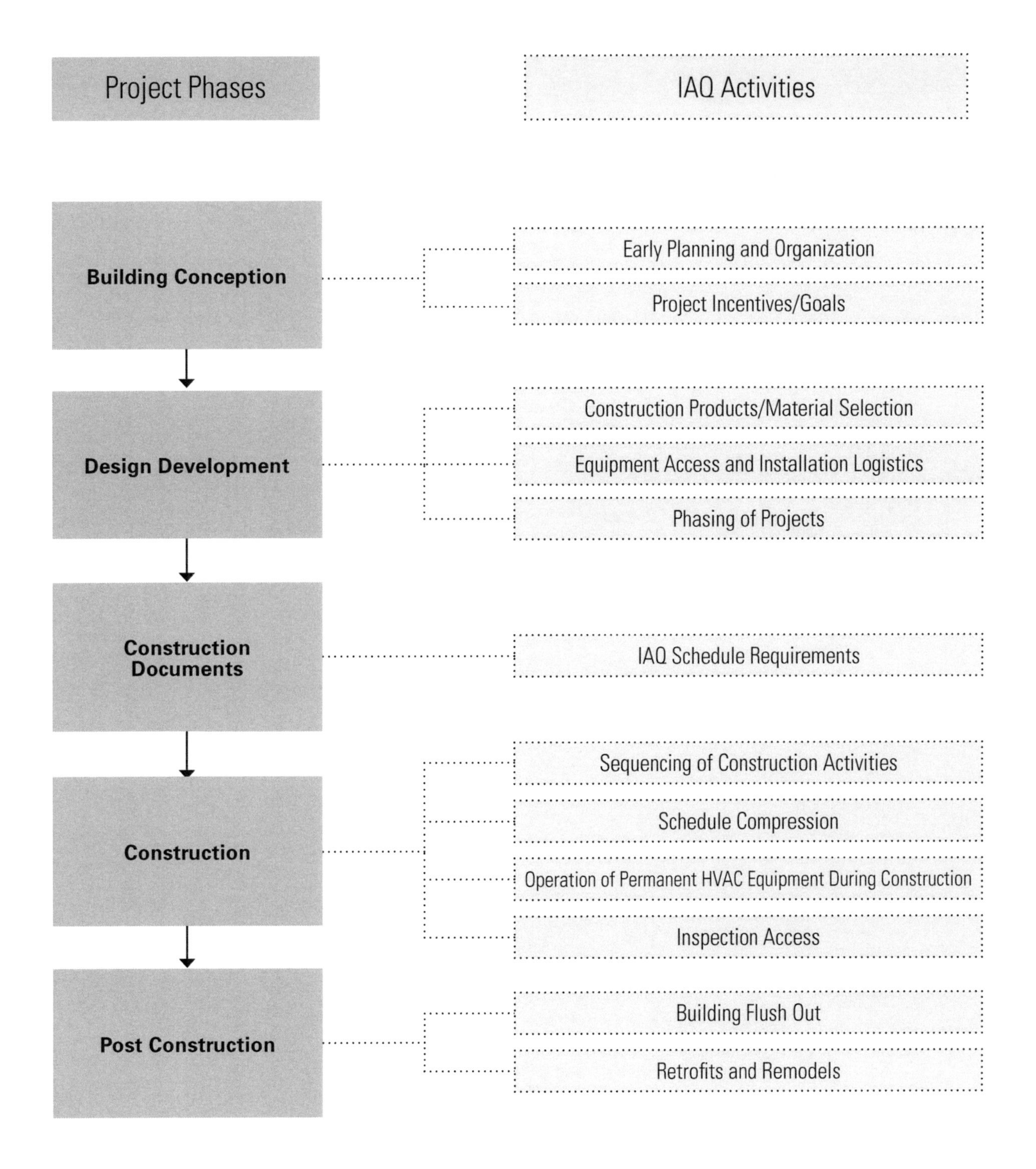

## Improper Storage and Installation of Construction Materials

Figures 1.4-A through 1.4-D illustrate a case of improperly storing and installing water-sensitive construction materials. The drywall in this project was installed after exposure to the elements. Before installation, the drywall had been stored outside and exposed to the weather and not properly protected. This allowed for moisture to invade the drywall and create an opportunity for microbial growth.

**Figure 1.4-A** Unprotected Storage of Water-Sensitive Materials

**Figure 1.4-B** Installation of Water-Sensitive Materials in an Exposed/Unprotected Structure

**Figure 1.4-C** Mold Growth from Wetting of Exposed Water-Sensitive Materials—Example 1

**Figure 1.4-D** Mold Growth from Wetting of Exposed Water-Sensitive Materials—Example 2

*Photographs courtesy of George DuBose.*

Nov
10
1.4

# Facilitate Effective Operation and Maintenance for IAQ



Operation and maintenance (O&M) can have as great an effect on IAQ as design and construction. O&M itself is outside the scope of this Guide, but there are steps the project team can take that greatly enhance the potential for effective O&M after the building is turned over to the owner. Among the most important are the following:

- Consider the owner's expected level of O&M capability when selecting systems.
- Involve O&M staff during project planning, design, construction, and Cx whenever possible.
- Provide O&M documentation that explains the design intent of key systems and how they need to be operated and maintained to fulfill that intent.
- Provide truly substantive training of O&M staff, emphasizing how systems need to be operated and maintained to achieve their design intent.
- Prioritize IAQ-related O&M documentation and training to emphasize the issues most important to IAQ for a given project.

O&M staff size, skill level, and budget need to be considered in selecting and designing systems. When systems are too complex or require maintenance that is too frequent or requires more advanced skills than the resources available, they will not be properly operated and maintained. When O&M resources are limited, selecting systems that are simpler, more forgiving, less numerous, and less difficult or time consuming to access is an important aspect of design for IAQ.

Involving O&M staff during planning, design, construction, and Cx facilitates effective O&M after turnover for two reasons. First, O&M staff can provide valuable input on issues that will affect operability and maintainability, such as standardization of equipment and components to reduce O&M costs, experience with product quality and vendor responsiveness, O&M training needs, and other issues. Second, staff can gain knowledge about the design and construction that will be useful in operating and maintaining the building.

Providing system-oriented O&M documentation, in addition to the customary component-oriented information, facilitates more effective O&M long after project completion. O&M documentation ought to explicitly communicate the purpose and intended performance of key IAQ-related building systems. If O&M staff do not understand the design intent of these systems, they will be much less able—or indeed motivated—to operate and maintain the systems to fulfill that intent. ASHRAE Standard 62.1 (ASHRAE 2007a), establishes minimum requirements for documentation of the design intent of ventilation systems that can serve as a starting point for compiling system-oriented O&M information. But design intent documentation also needs to cover other aspects of ventilation system design, other mechanical systems, and the building enclosure assemblies that can have a significant impact on IAQ. *ASHRAE Guideline 1.1-2007, HVAC&R Technical Requirements for The Commissioning Process* (ASHRAE 2007b), and *NIBS Guideline 3-2006, Exterior Enclosure Technical Requirements for the Commissioning Process* (NIBS 2006), provide useful guidance and examples of design intent documentation for HVAC systems and building exterior enclosures, respectively.

When projects are commissioned, the Cx reports need to be included in the O&M documentation. These reports document the conformance of systems and equipment to the design intent at turnover, providing a benchmark against which later performance can be compared. They also provide inspection checklists and test procedures that can be used as is or adapted for later inspection and testing.

# Strategy 1.5

**Involving O&M Staff in Planning, Design, Construction, and Commissioning**

**Considering O&M Capabilities in System Selection**

**Providing O&M Documentation that Facilitates Delivery of the Design Intent**

Owner's Project Requirements and Basis of Design

Record Documents

Commissioning Report

Operations Manual

Training Manual

Maintenance Manual

Format of O&M Documentation

**Prioritizing O&M for IAQ**

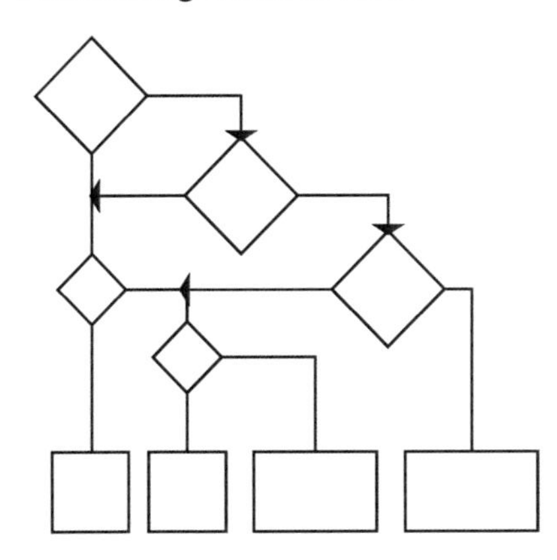

**Providing O&M Training to Support Delivery of the Design Intent**

The operations manual needs to include system-oriented information that can be used to proactively manage building operation. This includes standards of performance for relevant IAQ parameters such as space humidity and temperature, ventilation rate, and building pressurization, as well as log forms for recording periodic measurements of these parameters. Other useful elements include thorough system descriptions, operating routines and procedures, and seasonal start-up and shutdown procedures.

The maintenance manual needs to provide recommended maintenance activities and intervals for key IAQ-related systems. ASHRAE Standard 62.1 establishes minimum requirements for the maintenance of ventilation systems. *ANSI/ASHRAE/ACCA Standard 180, Standard Practice for Inspection and Maintenance of Commercial Building HVAC Systems* (ASHRAE 2008b) and the U.S. Environmental Protection Agency (EPA) *IAQ Building Education and Assessment Mode* (I-BEAM; EPA 2008a) and IAQ Tools for Schools Program (EPA 2007) provide additional guidance.

Providing O&M documentation in electronic as well as printed form makes it easier for staff to retain and update the information, make it available to multiple users, and search it for specific information.

O&M training needs to explain the design intent of key systems, their intended performance, and how they need to be operated and maintained to achieve that performance. Training materials need to be included in the O&M documentation and need to provide enough information so that a new operator using them for self-directed study can understand how to operate the systems properly. Recording O&M training sessions makes the training available for later reference by the same or new O&M staff.

Commissioning (see Strategy 1.2 – Commission to Ensure that the Owner's IAQ Requirements are Met) is an effective process to address O&M documentation and training needs. Other aspects of project design and execution also influence operability and maintainability for IAQ. These are discussed as separate Strategies, including, for example:

- Strategy 2.5 – Select Suitable Materials, Equipment, and Assemblies for Unavoidably Wet Areas
- Strategy 3.5 – Provide Effective Track-Off Systems at Entrances
- Strategy 3.6 – Design and Build to Exclude Pests
- Strategy 4.3 – Facilitate Access to HVAC Systems for Inspection, Cleaning, and Maintenance
- Strategy 4.4 – Control Legionella in Water Systems
- Strategy 5.3 – Minimize IAQ Impacts Associated with Cleaning and Maintenance

# Including the Ventilation Design Intent in Project Drawings Facilitates Proper Building Operation

1.5

**Figure 1.5-A** Minnesota School Building
*Photograph courtesy of Center for Energy and Environment.*

A Minnesota elementary school (Figure 1.5-A) had many of its air handlers replaced to bring ventilation rates up to current standards. Figures 1.5-B and 1.5-C show a portion of the design intent documentation provided by the mechanical engineering firm. This firm includes these tables as part of the drawings, on a sheet immediately following the usual HVAC equipment schedules. Since plans are often the longest-lived part of the building documentation, putting these tables in the plans helps to ensure that the ventilation system design intent is available for O&M staff over the long term.

Including this information as part of the O&M documentation enables facility staff to understand the rationale for minimum outdoor airflow rates. The documentation and training also needs to describe how these flow rates are to be set or controlled for each air-handling system. The design intent data can be used to develop operating standards and logs for inclusion in the operations manual. This clearly communicates the owner's building performance standards to O&M staff and fosters accountability for building IAQ performance.

| VUV - Typical Classroom (28 VUV's Total) | | | | | | | | | | | |
|---|---|---|---|---|---|---|---|---|---|---|---|
| zone name | zone ceiling height | zone area A (z) | zone people P (z) | oa rate/ person R (p) | oa rate/ area R (a) | zone cfm V (pz) | cfm per sq ft | a/ch per hour | breathing zone oa V (bz) | air dist. zone eff. E (z) | design zone oa V (oz) |
| classroom | 9 | 1,000 | 30 | 10 | 0.12 | 1,250 | 1.25 | 8.3 | 420 | 0.8 | 525 |
| totals | | 1,000 | 30 | | | 1,250 | | | 420 | | 525 |
| design results | | | | | | | | | | | |
| system total supply cfm | | 1,250 | | | | | | | | | |
| system total OA cfm V (ot) | | 525 | 42% | | | | | | | | |

**Figure 1.5-B** Ventilation Design Intent Data for Vertical Unit Ventilators Serving Classrooms
*Adapted from Hallberg Engineering, Inc.*

| RTU - 3 Design Intent Schedule | | | | | | | | | | | |
|---|---|---|---|---|---|---|---|---|---|---|---|
| zone name | zone ceiling height | zone area a (z) | zone people p (z) | oa rate/ person r (p) | oa rate/ area r (a) | zone cfm v (pz) | cfm per sq ft | a/ch per hour | breathing zone oa v (bz) | air dist. zone eff. e (z) | design zone oa v (oz) |
| 1301 computer lab | 9 | 940 | 20 | 10 | 0.12 | 1,950 | 2.07 | 13.8 | 313 | 0.8 | 391 |
| 1301b computer lab | 9 | 385 | 10 | 10 | 0.12 | 840 | 2.18 | 14.5 | 146 | 0.8 | 183 |
| 1301a office | 9 | 150 | 1 | 5 | 0.06 | 210 | 1.40 | 9.3 | 14 | 0.8 | 18 |
| totals | | 1,475 | 31 | | | 3,000 | | | 473 | | 591 |
| design results | | | | | | | | | | | |
| system total supply cfm | | 3,000 | | | | | | | | | |
| system total oa cfm v (ot) | | 591 | 20% | | | | | | | | |

**Figure 1.5-C** Ventilation Design Intent Data for a Rooftop Unit Serving Computer Labs
*Adapted from Hallberg Engineering, Inc.*

# Control Moisture in Building Assemblies

Moisture is one of the most common causes of IAQ problems in buildings and has been responsible for some of the most costly IAQ litigation and remediation. Moisture enables growth of microorganisms, production of microbial VOCs and allergens, deterioration of materials, and other processes detrimental to IAQ. In addition, dampness has been shown to be strongly associated with adverse health outcomes. Control of moisture is thus critical to good IAQ.

- Penetration of rainwater or snowmelt into the building envelope is a common cause of IAQ problems. Strategy 2.1 – Limit Penetration of Liquid Water into the Building Envelope describes design features and quality control processes that can limit water entry.
- Condensation is another common cause of IAQ problems. It most often occurs when moist air infiltrates into or exfiltrates out of the building enclosure and encounters a surface with a temperature below the air dew point. However, it can also occur due to vapor diffusion, capillary transport, or thermal bridging. Strategy 2.2 – Limit Condensation of Water Vapor within the Building Envelope and on Interior Surfaces describes design and quality control to reduce the likelihood of condensation problems.
- Negative building pressure can draw moist outdoor air into the building envelope, potentially leading to condensation. It can also draw moist air into the conditioned space itself, potentially increasing the latent load beyond the cooling system design capacity and leading to elevated indoor humidity. Positive building pressure can push moist indoor air into the building enclosure, potentially leading to condensation under heating conditions. Strategy 2.3 – Maintain Proper Building Pressurization addresses pressurization control.
- High indoor humidity increases the risk of microbial growth and IAQ problems. Strategy 2.4 – Control Indoor Humidity addresses humidity control, especially in hot, humid climates where controlling indoor humidity can be particularly challenging.
- Some indoor areas, such as shower rooms, toilet rooms, janitorial closets, and kitchens, frequently are wetted with liquid water or experience condensation due to high humidity. Strategy 2.5 – Select Suitable Materials, Equipment, and Assemblies for Unavoidably Wet Areas describes strategies to preserve IAQ in wet areas.
- Strategy 2.6 – Consider Impacts of Landscaping and Indoor Plants on Moisture and Contaminant Levels provides information on the advantages and disadvantages of plants from an IAQ perspective.

Other important moisture control issues are discussed in the following sections:

- Strategy 1.4 – Employ Project Scheduling and Manage Construction Activities to Facilitate Good IAQ
- Objective 4 – Control Moisture and Contaminants Related to Mechanical Systems

# Objective 2

**2.1** Limit Penetration of Liquid Water into the Building Envelope

**2.2** Limit Condensation of Water Vapor within the Building Envelope and on Interior Surfaces

**2.3** Maintain Proper Building Pressurization

**2.4** Control Indoor Humidity

**2.5** Select Suitable Materials, Equipment, and Assemblies for Unavoidably Wet Areas

**2.6** Consider Impacts of Landscaping and Indoor Plants on Moisture and Contaminant Levels



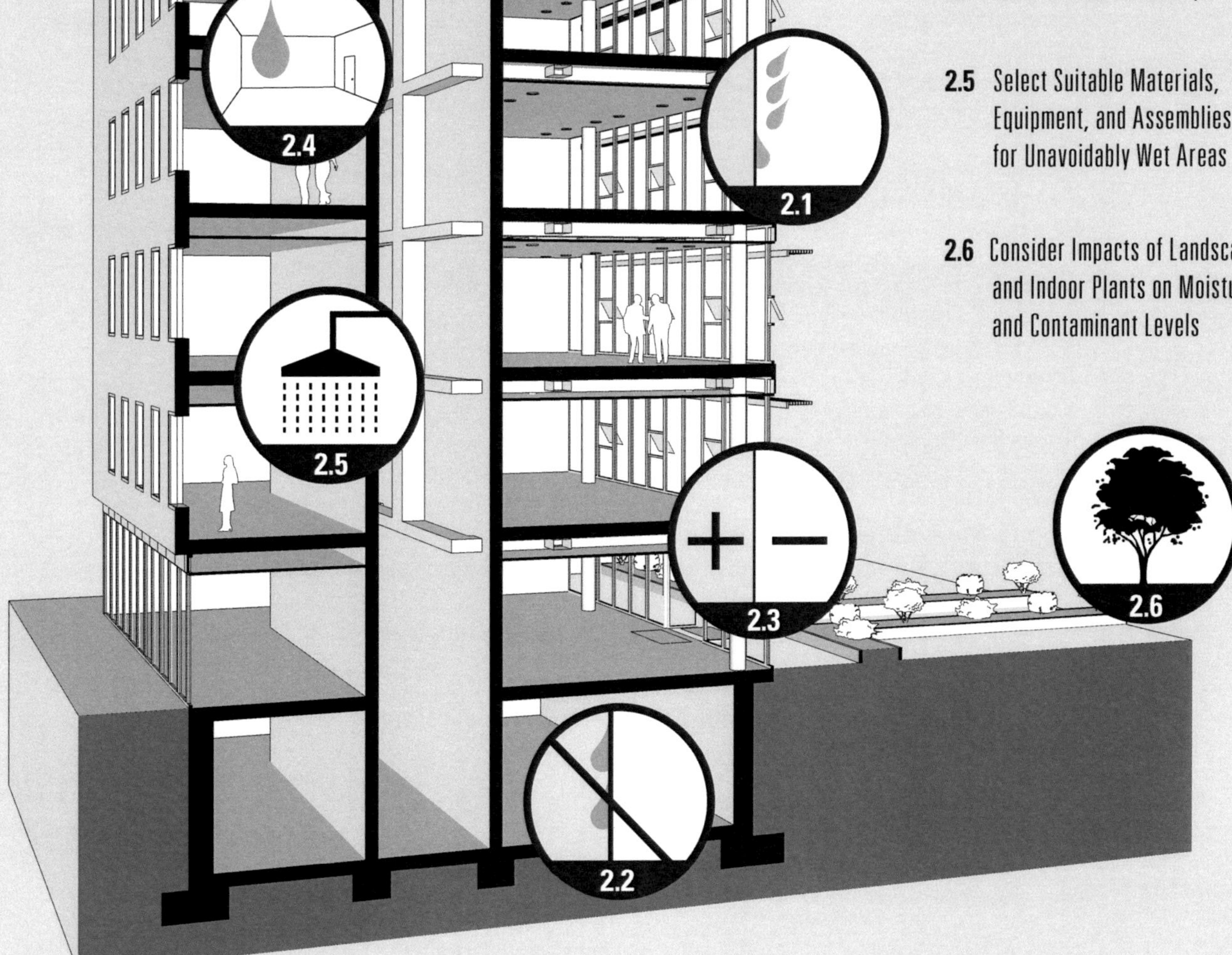

# Limit Penetration of Liquid Water into the Building Envelope



Moisture in buildings is a major contributor to mold growth and the poor IAQ that can result. Wetting of building walls and rainwater leaks are major causes of water infiltration. Preventive and remedial measures include rainwater tight detail design, selection of building materials with appropriate water transmission characteristics, and proper field workmanship quality control.

Effective liquid water intrusion control requires both of the following:

- Barriers to water entry established and maintained using capillary and surface tension breaks in the building enclosure.
- Precipitation shed away from the building using continuous effective site drainage and a storm water runoff system.

**Establish and Maintain Barriers to Water Entry**

Leaking rainwater can cause great damage to a building and the materials inside. Rainwater that falls on the building is controlled by a combination of drainage and capillary breaks. A capillary break keeps rainwater from wicking through porous materials or through cracks between materials and thus entering the building. Creating a capillary break involves installing a material, such as rubber roofing, that does not absorb liquid water. Another way to create a capillary break is to provide an air gap between materials that get wet and materials that should stay dry. An example of an air gap is the space behind brick veneers in exterior walls. Wall systems must employ cladding and flashing systems that direct the water away from the building.

The moisture-resistant materials that form the exterior skin of a building intercept and drain rain from roofs and away from walls, down walls and over windows and doors, and away from foundations (above, at, and below grade).

Sometimes a single moisture-impermeable material, sealed at the seams, forms the entire rainwater barrier—drainage and capillary break all in one. Membrane roofing and some glass panel claddings work in this way. Usually, however, roofing and cladding systems are backed up by an inner layer of moisture-resistant material that forms the drainage plane. The drainage plane intercepts rainwater that seeps, wicks, or is blown past the outer layer and drains it out of the building. An air gap between the drainage plane and the roofing or cladding provides a channel for drainage. The air gap and the drainage plane form capillary breaks between the outer layer and the materials inboard of the drainage plane.

**Directing Water Away from the Building**

The first step in rainwater control is to effectively situate the building and use or change the landscape to divert rainwater away from the structure. These actions are known as *site drainage* and include sloping the grade away from the building to control surface water and diverting water from the foundation below grade.

Once the site is designed properly to drain water away from the building, the building needs a storm water runoff system to divert precipitation from the roof into the site drainage system. This component of moisture control is called *storm water runoff management*.

The building foundation needs to be detailed to protect the building from rainwater. The above-grade portions of a foundation are often heavy masonry or concrete walls. A great deal of the rainwater that wets the above-grade wall simply drains off the surface to the soil below. Masonry and concrete walls are so massive, absorbed water is more likely to be stored in the wall—drying out between storms—than to wick through to the interior.

Landscape surfaces immediately surrounding the foundation perform the same function for the walls below grade as the roofing and cladding in the walls above grade: they intercept and drain rain away from

# Strategy 2.1

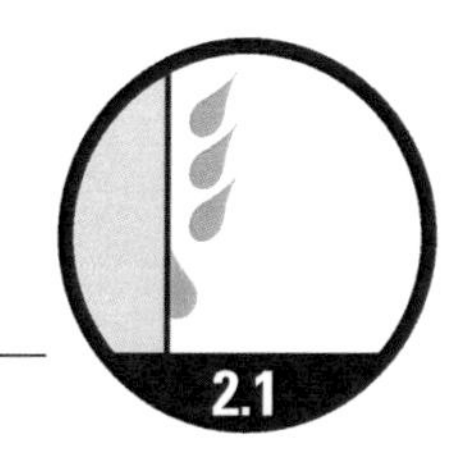
2.1



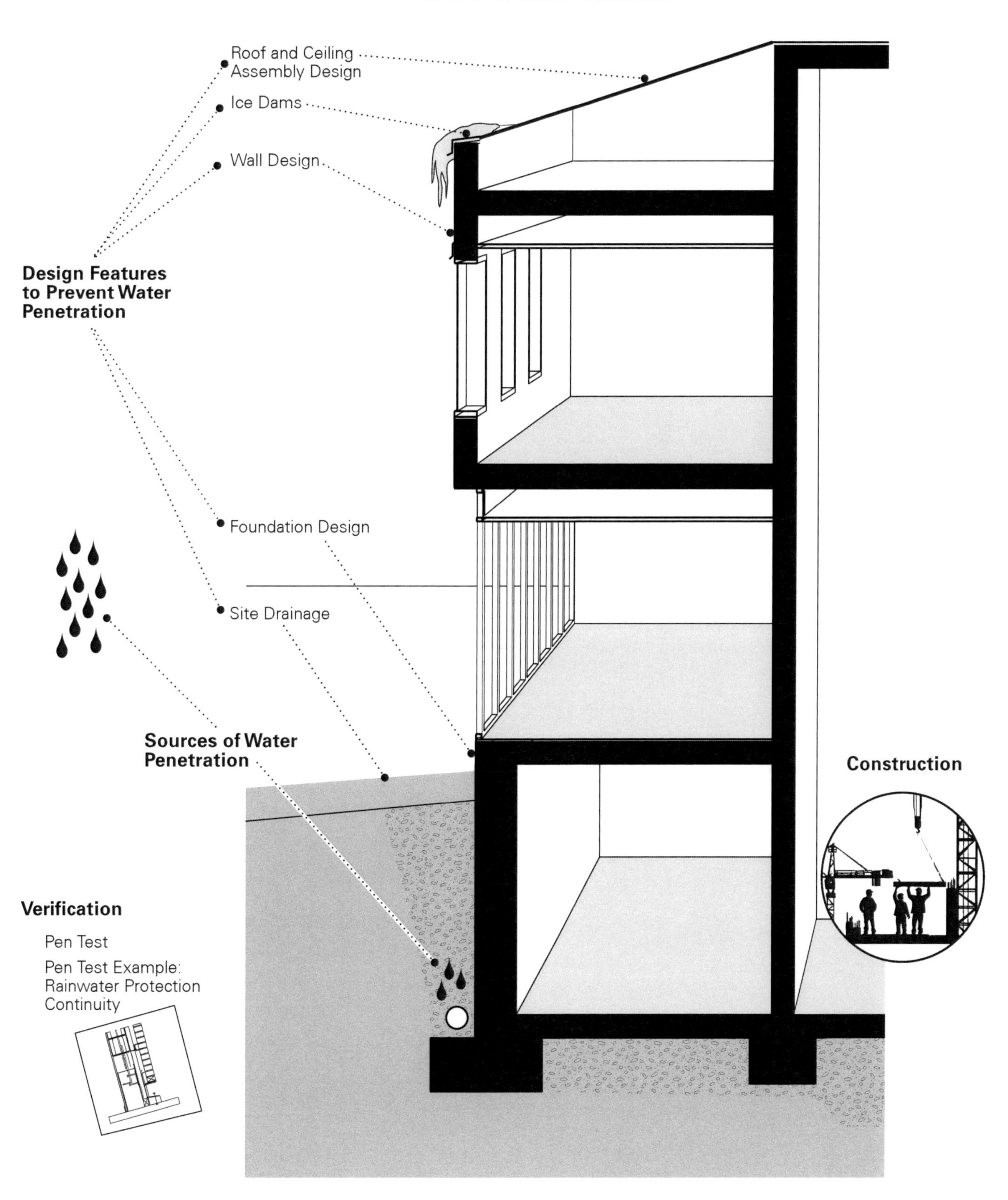
Sources of Water Penetration
Roof and Ceiling Assembly Design
Ice Dams
Wall Design
Design Features to Prevent Water Penetration
Foundation Design
Site Drainage
Sources of Water Penetration
Construction
Verification
Pen Test
Pen Test Example: Rainwater Protection Continuity

the building. The dampproof or waterproof coatings on below-grade walls serve the same purpose as the drainage plane in the above-grade walls, presenting a capillary break for rainwater that infiltrates the surrounding fill. Free-draining fill or geotechnic drainage mats placed against the below-grade walls serve the same function as the air gap in the above-grade walls; they provide a place for water to run down the drainage plane.

At the bottom of the below-grade wall, a footing drain system diverts rainwater or, in some cases, rising groundwater from the footing and the floor slab. Paint designed for use on concrete can be used on top of the footing to provide a capillary break between the damp footing and the foundation wall.

A layer of clean, coarse aggregate with no fines can provide a capillary break between the earth and the concrete floor slab. Plastic film beneath the floor slab provides a code-required vapor barrier and a capillary break beneath the slab.

## Water Intrusion in a Multi-Family Complex

This multi-family complex (Figure 2.1-A) had several areas of water intrusion through the building envelope. These areas included the roof to vertical wall intersection, the window and window surround, the penetrations, and the elevated slabs. The water intrusion resulted in damage in these areas and the need to remove portions of the building envelope to repair the damage. The cost of the remediation was estimated to be over $2 million.

The areas of failure included penetrations that were not flashed or sealed as well as windows and other openings with improper flashing around them. Damage to these areas because of these failures required remediation of the building envelope to correct them:

- Penetrations through the waterproofing membrane resulted in a breach in the capillary plane and deterioration of the underlying structure (Figure 2.1-B).

**Figure 2.1-A** Points of Water Entry into the Building Envelope

- Lack of flashing and sealant at vent penetrations through the veneer resulted in water intrusion (Figure 2.1-C).
- Lack of flashing at the windows for control of water drainage at the window openings resulted in intrusion (Figure 2.1-D).
- Lack of complete flashing at the low roof intersection and the adjacent vertical wall (rake wall condition) resulted in water intrusion into the wall (Figure 2.1-E).

**Figure 2.1-B** Structure Damage from Failures in the Building Envelope

**Figure 2.1-C** Damage from Failure at Vent Penetration

**Figure 2.1-D** Damage from Window Surround and Window Flashing Failures

**Figure 2.1-E** Damage from Failure of the Roof Flashing Termination

# Limit Condensation of Water Vapor within the Building Envelope and on Interior Surfaces



Preventing condensation or, more accurately, controlling the moisture content of building materials helps prevent the growth of microorganisms, especially molds, within the enclosure and on interior surfaces. The growth or amplification of microorganisms in buildings not only results in biodeterioration of susceptible construction materials but also leads to the production of allergens and microbial VOCs (which cause musty odors) that can affect occupant health and air quality. A complex microbial ecology can develop in or on construction materials that are chronically wet or damp (e.g., mites feed on mold; other organisms feed on mites). Allergens associated with molds and arthropods growing in chronically wet construction niches can enter the indoor environment and pose a risk to sensitive occupants.

The moisture content of building materials increases due to water vapor transport across enclosure assemblies either due to infiltrating, exfiltrating, or convecting air in contact with surfaces that have a temperature lower than the dew point of the air coming in contact with the surface and/or by diffusion due to a difference in water vapor pressure across the assembly or by capillary transport through the microscopic voids in building materials. Thermal bridges in the form of highly conductive materials that penetrate the insulated enclosure can drop temperatures of indoor surfaces to levels promoting condensation. Properly designed enclosure assemblies that have greater drying potential than wetting potential and that achieve a moisture balance over time are not always implemented, and many building designs do not get scrutinized for appropriate enclosure design.

Building enclosures are often designed without a proper understanding of the performance of the assembly when it is subjected to the exterior weather and interior boundary conditions. Code requirements may even impose solutions that are problematic, such as requiring vapor retarders prescriptively or requiring water-resistive barriers that may be too vapor permeable under certain conditions. Prescriptive criteria in codes are slowly being improved, but the substitution of a single material in an assembly can radically change how the assembly performs over time.

Building enclosures need to be designed by a knowledgeable design professional using design tools referenced in *ASHRAE Handbook—Fundamentals* (ASHRAE 2009) in order to avoid the likelihood of moisture-related problems.

**Design for Airtightness of the Enclosure**

A continuous air barrier system in the building enclosure needs to be included. A continuous air barrier system is created by adhering to the following steps.

- Select a material in each opaque wall, floor, and roof assembly that meets a maximum air permeance of 0.004 cfm/ft$^2$ at 0.3 in. w.g. (0.02 L/s·m$^2$ at 75 Pa) and join it together with tapes, sealants, etc., into an assembly.
- Join the air barrier layer of each assembly with the air barrier layer of adjacent ones and to all fenestration and doors until all enclosure assemblies (for the complete building as a six-sided box) are interconnected and sealed.
- Seal all penetrations of the air barrier layer. The airtight layer of each assembly will support the entire air pressure caused by wind, stack effect, and HVAC operation.

# Strategy 2.2

2.2



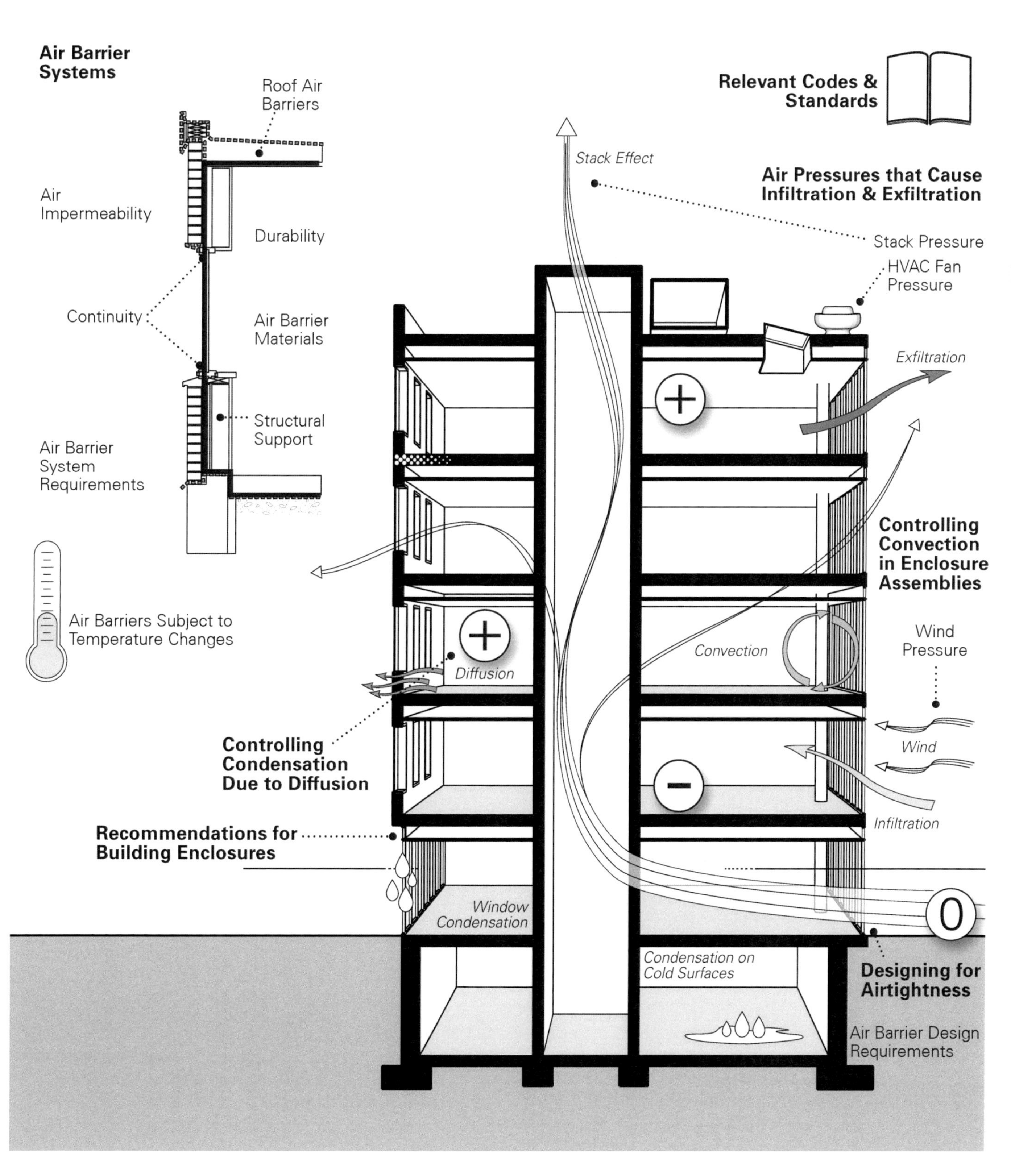
Air Barrier Systems
Roof Air Barriers
Air Impermeability
Durability
Continuity
Air Barrier Materials
Structural Support
Air Barrier System Requirements
Air Barriers Subject to Temperature Changes
Relevant Codes & Standards
Stack Effect
Air Pressures that Cause Infiltration & Exfiltration
Stack Pressure
HVAC Fan Pressure
Exfiltration
Controlling Convection in Enclosure Assemblies
Convection
Diffusion
Wind Pressure
Wind
Controlling Condensation Due to Diffusion
Infiltration
Recommendations for Building Enclosures
Window Condensation
Condensation on Cold Surfaces
Designing for Airtightness
Air Barrier Design Requirements

- Ensure that the airtight layer is structurally supported and can support the maximum positive and negative air pressures it will experience without rupture, displacement, or mechanical damage. Stresses must be safely transferred to the structure.

### Design for Convection

Air gaps adjacent to cool or cold surfaces can promote convection within a wall assembly. Cold air is heavier and drops, pulling in warm humid air to replace it and deposit moisture on the cold surface. This is especially true in vertical or sloping assemblies. The colder side can be the sheathing in colder climates or the interior drywall in warmer climates. Eliminating the air space on one or the other side of the insulation can be effective in preventing these convective loops. Fibrous insulation, however, which is mostly air, can also promote these convective loops.

### Design for Diffusion

A vapor retarder with appropriate permeance for the application should be placed on the predominantly high vapor pressure side of the assembly. To design assemblies for appropriate diffusion control, hygrothermal analysis is needed using either the steady-state dew point or the Glaser method, described in Chapter 25 of *ASHRAE Handbook—Fundamentals* (ASHRAE 2009) or by using a mathematical model that simulates transient hygrothermal conditions (such as WUFI, hygIRC, or Delphin). Users of such methods need to understand their limitations, and interpretation of the analysis results should be done by a trained person to reasonably extrapolate field performance approaching the design results. The International Energy Agency Annex 14 (IEA 1991) has established that a surface humidity of 80% represents a reasonable threshold for designers to achieve a successful building enclosure assembly for temperatures between 40°F and 120°F (5°C and 50°C).

### Window and Skylight Selection

Fenestration should be selected carefully by designers to avoid condensation. Fenestration is selected taking into account the interior boundary conditions and exterior weather conditions, and, from a chart developed by American Architectural Manufacturers Association (AAMA; 1988), the appropriate condensation resistance factor (CRF) for the window or skylight is determined. Thermally broken units that minimize the amount of exterior metal exposed to cold usually perform best. The edge spacer of the insulating glass unit is usually the most conductive (and coldest) location in an assembly. A new generation of "warm-edge" spacers that include thermally broken aluminum spacers, stainless steel spacers, and non-metallic glass-fiber reinforced plastic spacers are increasingly being used and improve the thermal performance of fenestration. Window and skylight manufacturers generally can provide National Fenestration Rating Council (NFRC) simulations using the software THERM (LBNL 2008) that show how a specific selection of window, spacer, and glass with various gaseous fill will perform. It is also important to note that some non-metal windows that have improved U-factors may have worse CRFs than metallic thermally broken windows. Custom designs are often required to be verified using the THERM and WINDOW (LBNL 2009a) software and validated by physical laboratory testing.

### Below-Grade Walls and Slabs on and Below Grade

Deep ground temperature in a locale is not unlike the average annual temperature, with local variations due to shading from vegetation, elevation, or proximity to the coast. Comparing the annual average temperature with the August dew-point temperature of the air is a good indication of whether mold will grow on slabs and walls of below-grade structures.

Concrete is highly conductive and its temperature will become very similar to ground temperature, making the concrete potentially become a condensation surface. Insulating outside the concrete is the best choice for keeping the concrete above the dew point of the air. In termite-infested areas, select rigid insulation that has termiticides included; this renders poisoning the soil unnecessary. Insulating under slabs with a vapor retarder on top in intimate contact with the slab is the best strategy for a dry slab. Insulating on the inside of below-grade walls is possible, but it is best to insulate using adhered rigid insulation so as to avoid convection through fibrous insulation.

Comparing a location's average annual temperature (Figure 2.2-A) with the August dew point of the air (Figure 2.2-B) is a good indication of whether below-grade structures may cause condensation and mold.

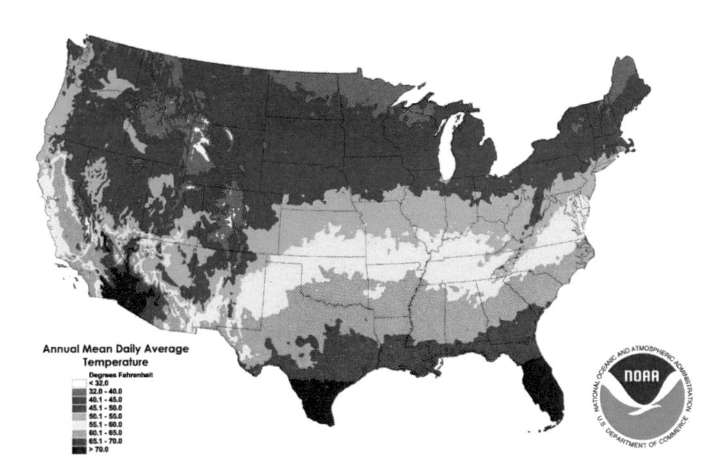


**Figure 2.2-A** Average Mean Daily Temperatures
*Image courtesy of National Climatic Data Center.*

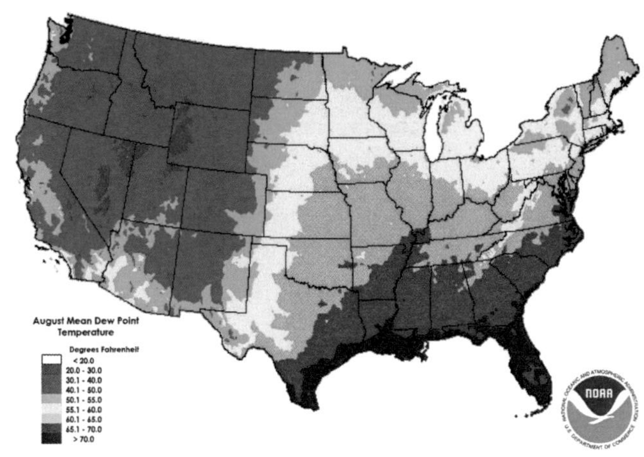


**Figure 2.2-B** Mean Dew-Point Temperatures for August
*Image courtesy of National Climatic Data Center.*

## Thermal Bridging

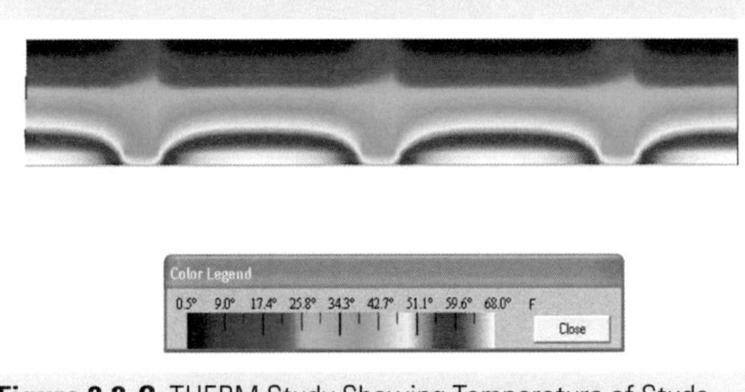


**Figure 2.2-C** THERM Study Showing Temperature of Studs

**Figure 2.2-D** Ghosting of Steel Studs on Walls
*Photograph copyright Wiss, Janney, Elstner and Associates, Inc.*

Thermal bridges, due to conductive materials that penetrate or interrupt the thermal insulation layer, cause a drop in temperature of the interior surface in cold climates (Figure 2.2-C). This can cause condensation and mold growth. It can also cause deposition of particulates onto the cold surfaces due to convective loops caused by the temperature differences, which is called "ghosting." In the building shown in Figure 2.2-D in a cold climate, interior humidity, candle smoke, and thermal bridging combined to cause ghosting of the steel studs.

# Maintain Proper Building Pressurization

Proper building pressurization is required to limit moisture and contaminant transfer across the building envelope. Moisture transfer can result in mold damage within the envelope and, along with other contaminant transfers, can contaminate occupied spaces within the building.



Building pressurization is the static pressure difference between the interior pressure and the exterior (atmospheric) pressure of a building. This static pressure difference influences how much and where exfiltration and infiltration occur through the building envelope. The static pressure difference across the envelope is not the same at all points of the building envelope. Wind direction and speed; indoor-outdoor temperature differences; differing mechanical supply, return, and exhaust airflows to each space; and compartmentalization of spaces can create different static pressures at various points of the building envelope. While many HVAC systems are designed to achieve an overall building pressurization of 0.02 to 0.07 in. w.c. (5 to 17 Pa) differential (across the building envelope) in the lobby, this is not always advisable. The nature and extent of the pressure differential will depend on a variety of factors that will need to be assessed. The actual pressure differential can fluctuate due to changing weather conditions, wind load, and HVAC system operation.

Positive building pressure is particularly important to maintain in the following situations: mechanically cooled buildings in hot and humid climates for reduced infiltration and control of condensation and mold growth, buildings maintained at low temperatures relative to outdoor temperatures (refrigerated warehouses, ice arenas, etc.) for reduced infiltration and control of condensation and mold growth, and buildings in areas with poor outdoor air quality to control the infiltration of the outdoor air contaminants.

Buildings in cold climates are typically designed for a neutral pressure to avoid exfiltration of relatively moist air during the heating season, which could cause condensation, mold in the building envelope, and deterioration of the building envelope. Similarly, humid spaces (e.g., natatoriums, shower rooms, spas, kitchens, indoor gardens, humidified buildings, or areas such as health-care facilities, museums, and musical instrument storage and performance areas) are of extra concern in cold climates and need to be slightly negative in pressure relative to the outdoors to reduce the risk of condensation and mold in the building envelope. A discussion on mixed climates can be found in the "Climatological Requirements" section in the Part II detailed guidance in the electronic version of this Guide.

Buildings are often treated as if they are one large compartment. In reality, buildings are typically composed of many smaller compartments or spaces. The static pressures differ from one space to another due to stack effect, changing wind direction, climate changes, HVAC system operation, etc. It is important that these factors be considered for proper compartmentalization and/or HVAC system control and system segregation. For example, maintaining positive building pressure in the lobby does not mean every space adjacent to the lobby is also positively pressurized. The top floors of a building in hot weather may be negatively pressured, and HVAC systems that employ return air plenum systems instead of return air ducts may have bands of negative pressure on each floor.

When designing for proper building pressurization, envelope leakage is often overlooked. The amount of envelope leakage can drastically change the required outdoor air (makeup air) quantity to maintain a positively pressured building.

# Strategy 2.3

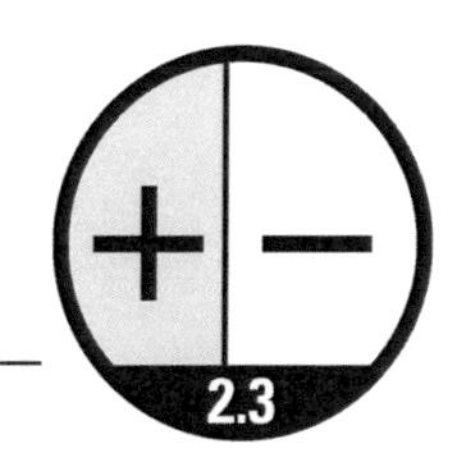

2.3




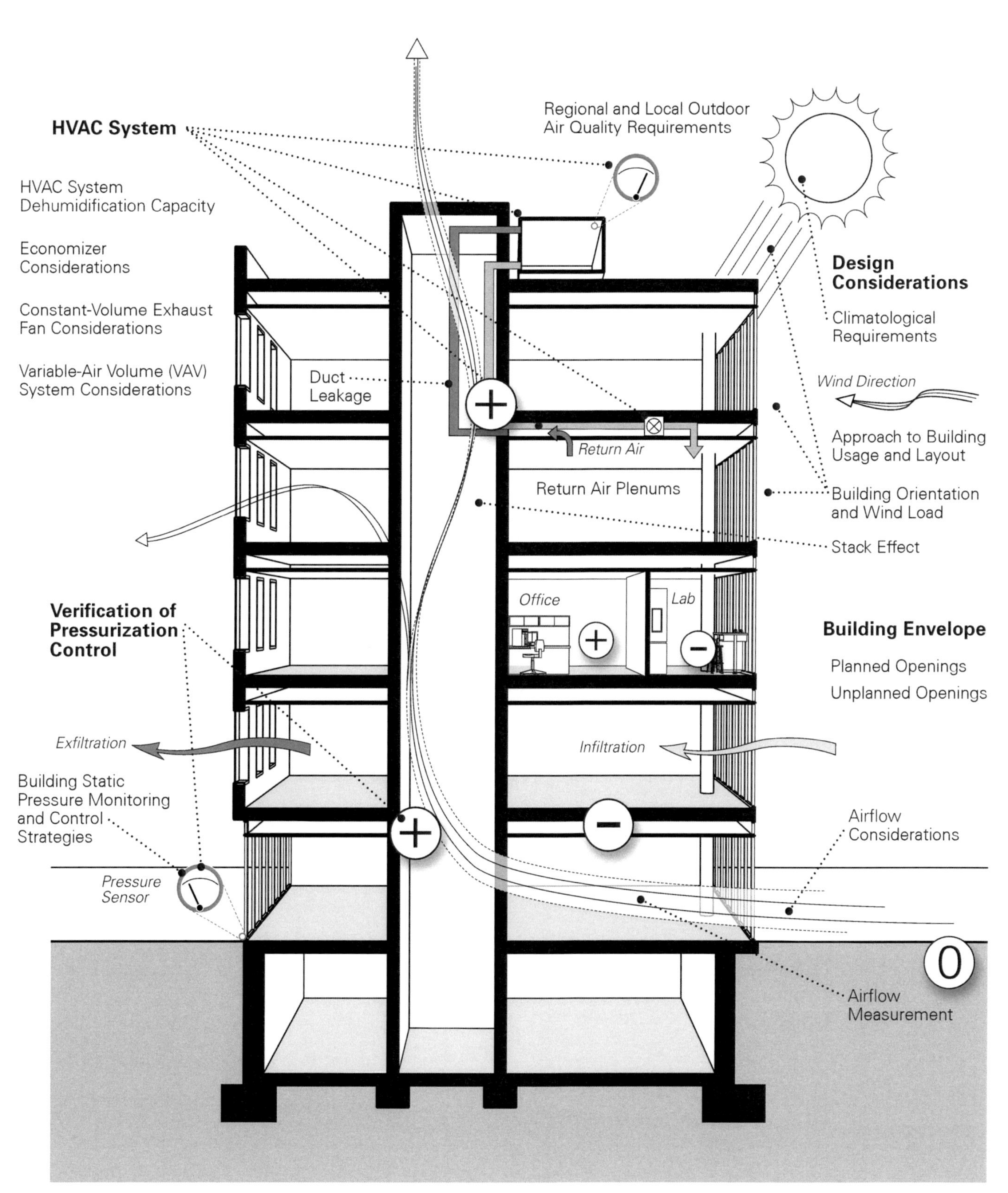

HVAC System
HVAC System Dehumidification Capacity
Economizer Considerations
Constant-Volume Exhaust Fan Considerations
Variable-Air Volume (VAV) System Considerations
Duct Leakage
Regional and Local Outdoor Air Quality Requirements
Design Considerations
Climatological Requirements
Wind Direction
Approach to Building Usage and Layout
Return Air
Return Air Plenums
Building Orientation and Wind Load
Stack Effect
Office
Lab
Verification of Pressurization Control
Building Envelope
Planned Openings
Unplanned Openings
Exfiltration
Infiltration
Building Static Pressure Monitoring and Control Strategies
Airflow Considerations
Pressure Sensor
Airflow Measurement

The simple assumption that outdoor air intake exceeds exhaust airflow typically does not ensure that the building will be positively pressurized. Envelope leakage, wind load, building size and dimensions, building orientation, compartmentalization, and building usage all need to be assessed. The more complex the building, including layout, building height, architectural features, HVAC system type and usage, etc., the more difficult proper building pressurization attainment will be.

In addition to the proper volume of air being provided for proper building pressurization, the distribution of the air within the building spaces needs to be addressed. Makeup air needs to be provided in the correct areas or spaces to help overcome depressurization due to stack effect and/or wind effect. See Strategy 7.3 – Effectively Distribute Ventilation Air to the Breathing Zone and Strategy 7.4 – Effectively Distribute Ventilation Air to Multiple Spaces for guidance on how this can be accomplished.

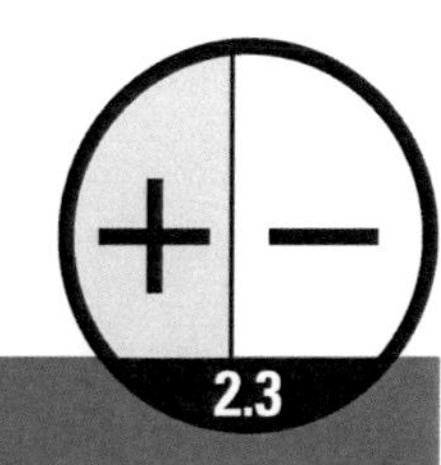


## Results of a 300-Room Building Negatively Pressurized

A 300-room building was negatively pressurized to the outdoors (Figure 2.3-A). The warm moist outdoor air infiltrated the building and traveled through the walls and sought entry points into each room. One of those points was the electrical outlets. The surfaces were cool enough (the rooms were air conditioned) to result in condensation and widespread mold growth throughout the facility, including on the furniture and in the walls of the building (Figure 2.3-B). Each room had individual exhaust with outdoor air introduced into the common corridor. Verification of the building pressure and individual room pressure never occurred. The building required complete renovation at a cost of $9.9 million.

**Figure 2.3-A** "Smoke" Test Demonstrating the Building was Negatively Pressurized in Reference to the Exterior

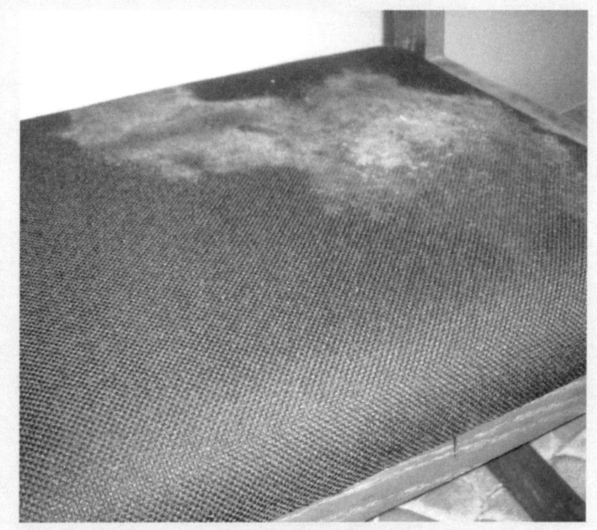

**Figure 2.3-B** Mold Growth due to Negative Pressure

*Photographs copyright Liberty Building Forensics Group®.*

STRATEGY OBJECTIVE 2.3

# Control Indoor Humidity

Control of indoor humidity is important for occupant health and comfort and because high humidity can cause condensation, leading to potential material degradation and biological contamination such as mold. High humidity also supports dust mite populations, which contribute to allergies. On the other hand, low humidity affects health by drying out mucous membranes. Humidity conditions also affect people's perception of IAQ. Also see Strategy 7.6 – Provide Comfort Conditions that Enhance Occupant Satisfaction.



Situations where special consideration should be given to humidity control include:

- hot and humid climates, especially when building pressurization is difficult to achieve or there are long periods of no conditioning (such as school systems shut down for the weekend);
- conditioned spaces with large indoor moisture sources;
- conditioned spaces with unusually cold surfaces;
- spaces with continuous outdoor air ventilation and non-continuous air conditioning (for instance, when cooling coils cycle on/off or exhaust fans must continue to run when conditioning systems are off); and
- oversized systems with excessive airflow with modulation of the chilled-water flow rate as the only available control method.

These are situations that contribute to the risk of localized excessive humidity and condensation in the presence of surfaces with temperatures below the dew point. In addition, any building in a cold climate may experience extremely low humidity, but this situation does not always mean that installation of a humidifier is advisable due to concerns about other potential problems.

### Principles of Condensation

As air is cooled, its capacity to hold moisture diminishes. When air cools enough that it becomes saturated (100% RH) so it can no longer hold all its water vapor, the vapor turns back into a liquid (condenses). The temperature at which this happens is a called the *dew point* and typically occurs on surfaces that are cooler than the dew point of the surrounding air.

### Integrated Design Process

Air-conditioning system designers often choose indoor conditions like 50% RH or 60% RH (ASHRAE Standard 62.1 requires 65% for systems that dehumidify [ASHRAE 2007a]) and design a system to handle the peak sensible load (i.e., peak dry-bulb temperature). However, most of the time systems operate at part load. Under these conditions, systems that control only space dry-bulb temperature may not provide enough dehumidification to keep space humidity within an acceptable range. For this reason, ASHRAE Standard 62.1 *now* requires that designers consider the dehumidification performance of the system at a "humidity challenge" condition intended to represent a part-load situation with high latent and low sensible load.

This change requires additional design effort for load calculations at more than one design condition, selection of automatic temperature control for humidity considerations, possibly a change of system type, and coordination with those selecting exterior walls and surfaces on the interior. Beyond these required

# Strategy 2.4

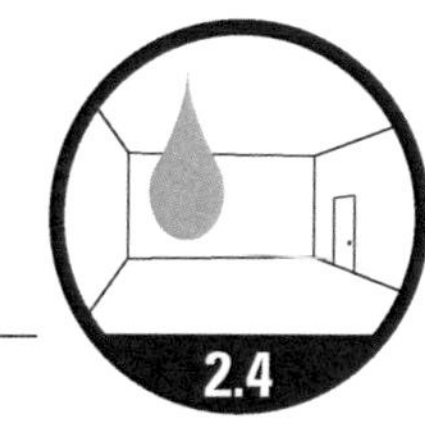
2.4

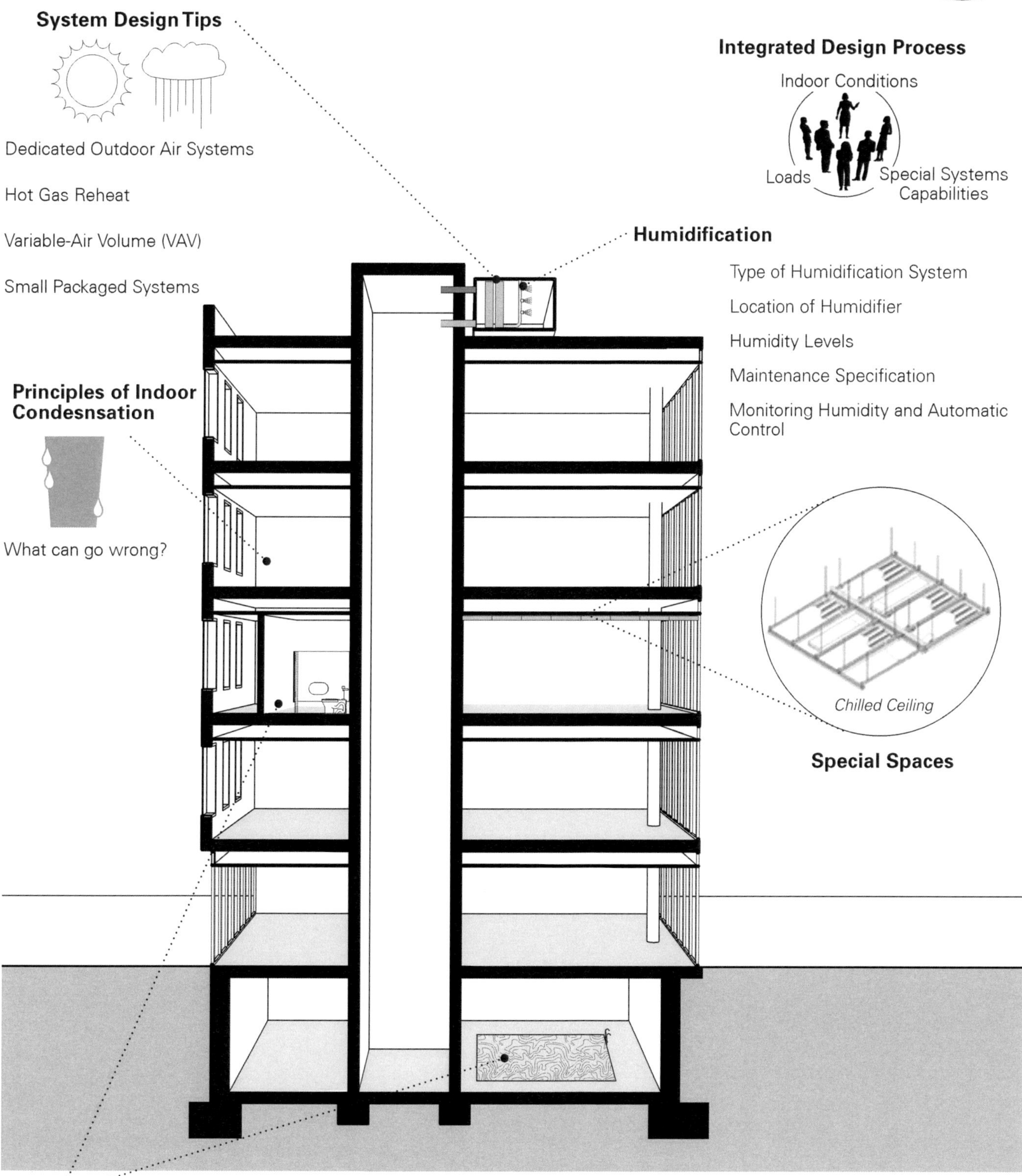
System Design Tips
Dedicated Outdoor Air Systems
Hot Gas Reheat
Variable-Air Volume (VAV)
Small Packaged Systems
Principles of Indoor Condesnsation
What can go wrong?
Integrated Design Process
Indoor Conditions
Loads
Special Systems Capabilities
Humidification
Type of Humidification System
Location of Humidifier
Humidity Levels
Maintenance Specification
Monitoring Humidity and Automatic Control
Chilled Ceiling
Special Spaces
Dedicated Dehumidification Systems

design considerations, IAQ would benefit from actual measurement of humidity (relative humidity or dew point) and feedback to control system parameters.

**System Design Tips**

In hot, humid climates or other situations with high latent loads, constant-volume systems with on/off cycling control may not provide adequate humidity control at part-load conditions. Cycling a direct expansion (DX) cooling system on and off to satisfy space temperatures or resetting the discharge air temperature upward reduces the system's ability to remove moisture from the supply air. Other system designs can keep indoor humidity within an acceptable range without the need for dehumidifiers or humidity control systems. These may include selecting a lower discharge air temperature (lower cfm/ton [L/s per kW] ratio) VAV control (even for single-zone systems), use of hot gas reheat, dedicated outdoor air systems (DOASs) and demand control of outdoor air.

Some of these strategies may not be available with smaller packaged cooling units but may be available with an upgrade of the HVAC equipment or a change of the system type. Packaged units may not have published selection data at low discharge air temperature (lower cfm/ton [L/s per kW] ratio), and the designer may need to contact the manufacturer. It may be necessary to put multiple spaces on a single packaged unit or built-up system in order to get access to features available on the larger units, such as lower discharge air temperature, compressor unloading, VAV systems, energy recovery, hot gas reheat, or demand control of outdoor air (for instance, by control of $CO_2$).

**Special Spaces**

Spaces that have large latent loads and small sensible loads, either at full load or at part load, may require dedicated humidity control systems. Some designs call for intentionally cool surfaces, such as a chilled ceiling system and uninsulated ductwork in occupied spaces. It is especially important to analyze the resulting space humidity to avoid condensation in these systems. A space that requires high outdoor air ventilation rates in humid climates is a candidate for energy recovery ventilation, primarily for the purpose of using less energy to cool and dehumidify the outdoor air.

**Dedicated Dehumidification Systems**

Dedicated humidity control systems (dehumidifiers) may be required in spaces that are underground, in swimming and bathing areas, in kitchens, or where large volumes of unconditioned humid outdoor air enters the space, for instance, by door openings or other forms of infiltration. Dehumidifiers may be based on the refrigerant cycle or use a desiccant. The latter is a material that is hygroscopic (attracts water) and removes water vapor from an airstream; it must be regenerated by heating to drive off the water as part of the operation cycle.

In showers, natatoriums, and cooking areas, it is accepted that humidity will be high and condensation will occur, and consequently surfaces need to be inorganic and cleanable (see Strategy 2.5 – Select Materials, Equipment, and Assemblies for Unavoidably Wet Areas). Moisture can be kept out of less moisture-tolerant parts of the building by keeping spaces with high humidity at lower pressure than adjoining spaces.

**Humidification**

Humidification of buildings may solve some comfort and health problems but may create others. For this reason, ASHRAE Standard 62.1 no longer requires a minimum humidity level in buildings, and many designers prefer to err on the side of no humidifier. However, this view is not universally held. It is possible that humidifiers can have real benefits if properly applied and maintained (Schoen 2006).

When designing a system with humidification, be aware of the requirements of ASHRAE Standard 62.1 and several additional design issues delineated in the following.

The ASHRAE Standard 62.1 (Section 5.13) requirements are the following:

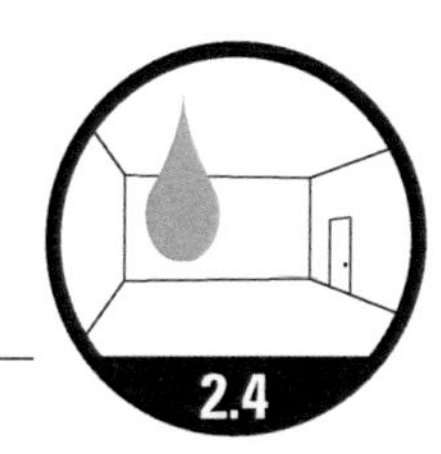


- Water must be from a potable or better source.
- Downstream air cleaners and duct obstructions such as turning vanes, volume dampers, and duct offsets should be kept greater than 15° away from the humidifier (as recommended by the manufacturer) or a drain pan should be provided to capture and remove water.

Additional design issues are the following:

- Consider preconditioning outdoor air using a total energy recovery wheel.
- Different types of humidifiers have different advantages and disadvantages. Avoid those using reservoirs of standing water and be aware of water treated with chemicals.
- Higher levels of indoor humidity concurrent with low outdoor temperatures increase the potential for condensation. Therefore, do not overhumidify. A setpoint of 20% RH or lower may be reasonable for many buildings during very cold weather.
- Locate the humidifier after the heating coil, where relative humidity is lowest, and provide water-tolerant, nonporous airstream surfaces downstream.
- Consult manufacturers and their representatives skilled in the application of systems in the climate and water conditions at the project location.
- Since the disadvantages of humidifiers are so driven by maintenance, it is especially important that building operators get additional assistance. Consider writing a maintenance specification for pricing with the installation.



**Monitoring Humidity and Automatic Control**

Whether the goal is to remove humidity from indoor air or to intentionally humidify, monitoring humidity is useful. Initially, monitoring can aid in verifying system performance during the test and balance/Cx process. During operation, monitoring can provide feedback and early warning of excursions.

Humidity is difficult to measure accurately, especially at very low or high relative humidity. Instruments to measure humidity cost significantly more than those for dry-bulb temperature and require more maintenance, and reliable standards against which to calibrate are more expensive. Inexpensive types of real-time monitoring that can prevent serious problems are the use of liquid moisture sensors in below-grade floors subject to flooding, secondary condensate pans, and water heater overflow pans.

## Assembly Room Humidity Control with VAV

An assembly room that is part of a large religious and educational facility has a single chilled-water air handler. The room is used for meetings, parties, and performances, and the chilled-water coil meets the full load, even at about 50% outdoor air. Conventional part-load controls would throttle the chilled-water valve at constant airflow, resulting in a warmer coil that would lose its ability to dehumidify unless reheat is applied. Figure 2.4-A and Table 2.4-A show how an upgrade of the air handler to VAV control accomplishes dehumidification without the use of reheat. The upgrade also includes DCV by sensing $CO_2$ and tracking the actual outdoor air supply using dedicated minimum outdoor air inlets with airflow sensing and dampers.

The psychrometric chart in Figure 2.4-A and the accompanying table (Table 2.4-A) represent the load conditions at the dehumidification design conditions and not the more common peak dry-bulb temperature. The system can meet the low room temperature and humidity design conditions. In order to do this, the coil discharge temperature is kept low (48.5°F [7.5°C]). The high latent load (54% of the total load) and the low airflow (about 180 cfm/ton [19 L/s per kW]) require a low entering chilled-water temperature (39°F [4°C]). In order to maintain humidity control, this particular design did not include upward reset of discharge air and chilled-water temperatures. Reset could be considered for efficiency but would require additional considerations for the control algorithm, such as possible measurement of room humidity as an input.

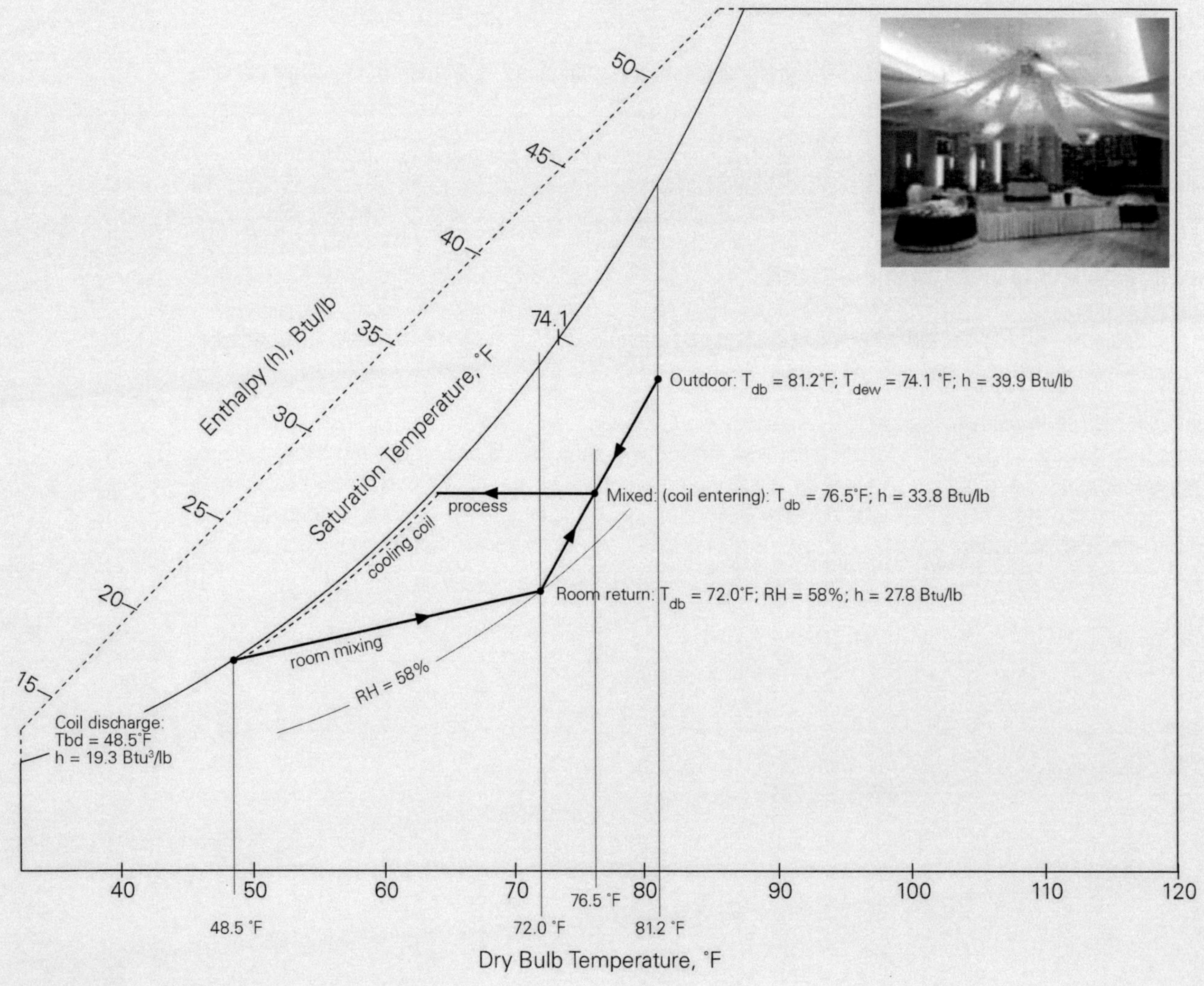


**Figure 2.4-A** Assembly Room Psychometric Chart
*Inset photograph courtesy of Larry Schoen.*

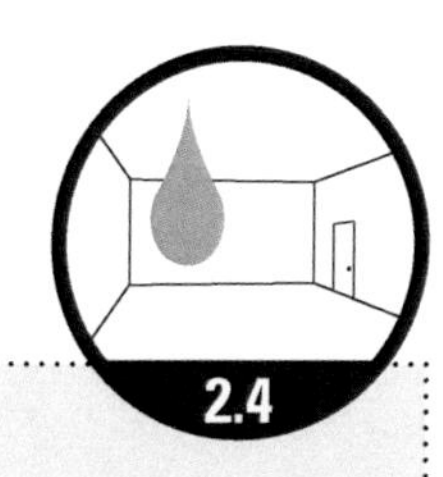


| Design Conditions | | |
|---|---|---|
| Room air temperature | 72.0°F (22.2°C) | Low design temperature for comfort |
| Room humidity | 58 % RH | Low design relative humidity for comfort |
| Outdoor air temperature at design dew point | 81.2°F (27.3°C) | Coincident dry bulb, not peak dry bulb |
| Outdoor air design dew point | 74.1°F (23.4°C) | This is the 1% design dew point |
| **New Fan Conditions** | | |
| Total airflow | 18345 cfm (8660 L/s) | Note the low airflow/ton (kW) |
| Outdoor air | 9000 cfm (4250 L/s) | Note the high percentage outdoor air |
| **New Cooling Coil Conditions** | | |
| Mixed/coil entering air temperature | 76.5°F (27.7°C) | |
| Coil leaving air temperature | 48.5°F (9.2°C) | Note the low temperature for dehumidification |
| Mixed/coil entering air enthalpy | 33.8 Btu/lb (60.7 kJ/kg) | |
| Leaving coil air enthalpy | 19.3 Btu/lb (27.0 kJ/kg) | |
| Total cooling | 1197 MBh (350 kW) | Corresponds to 100 tons (350 kW) at design dew point |
| Sensible cooing | 555 MBh (163 kW) | |
| Latent cooling | 642 MBh (187 kW) | 54% of the load is dehumidification |
| Coil sensible heat ratio (SHR) | 0.46 | Not the low coil SHR |
| Entering water temperature | 39.0°F (3.9°C) | Note the low chilled-water temperature required |
| Leaving water temperature | 49.0°F (9.4°C) | |
| Water flow | 239 gpm (15 L/s) | |

**Table 2.4-A** Assembly Room Dehumidification without Reheat

# Select Suitable Materials, Equipment, and Assemblies for Unavoidably Wet Areas



Certain areas within buildings will have high local humidity and/or the frequent presence of liquid water. These include washrooms, kitchens, and janitorial closets. Special attention must be paid to the material surfaces employed in these zones such that they can withstand the impact of frequent wetting. As stated by Latta in 1962, the harmful effects of water on building materials cannot be overemphasized, and the construction of durable buildings would be greatly simplified without water's influence (Latta 1962). More recently, the concept of the "4 D's" of moisture control in buildings was advanced: *deflection* (don't let the moisture in), *drainage* (give moisture a means of escape), *drying* (facilitate air movement/breathing to remove moisture), and *durability* (mold and corrosion resistance of materials susceptible to wetting) (CMHC 1998). These aspects are well studied when it comes to building envelope design (see Strategy 2.1 – Limit Penetration of Liquid Water into the Building Envelope, Strategy 2.2 – Limit Condensation of Water Vapor within the Building Envelope and on Interior Surfaces, and Strategy 2.3 – Maintain Proper Building Pressurization) but also apply to certain indoor spaces.

The best design efforts cannot prevent occasional wetting incidents, and certain indoor activities intentionally lead to damp conditions and materials. The main focus of this Strategy lies with the *durability* aspect of materials subjected to periodic wetting episodes within buildings. This Strategy identifies building areas that may be subject to repeated wetting, reviews the properties of materials that enable moisture resistance or tolerance, and finally describes those materials suitable for use in wet locations.

The growing concern regarding the impact of damp spaces on occupant health—specifically the involvement of damp materials in this regard—is highlighted in the recent report on Damp Indoor Spaces and Health (IOM 2004), which concludes that "studies should be conducted to evaluate the effect of the duration of moisture damage of materials and its possible influence on occupant health and to evaluate the effectiveness of various changes in building designs, construction methods, operation, and maintenance in reducing dampness problems" (p. 5). It is clear then that our knowledge in this area remains limited. Some practical advice can, however, be provided in terms of material selection. Materials that combine moisture-resistant and non-resistant layers (e.g., vinyl-coated wallboard) may be highly susceptible to mold growth when subjected to wetting. In zones with high humidity, the use of suspended ceilings can result in the creation of unconditioned spaces that may become susceptible to condensation.

# Strategy 2.5

2.5



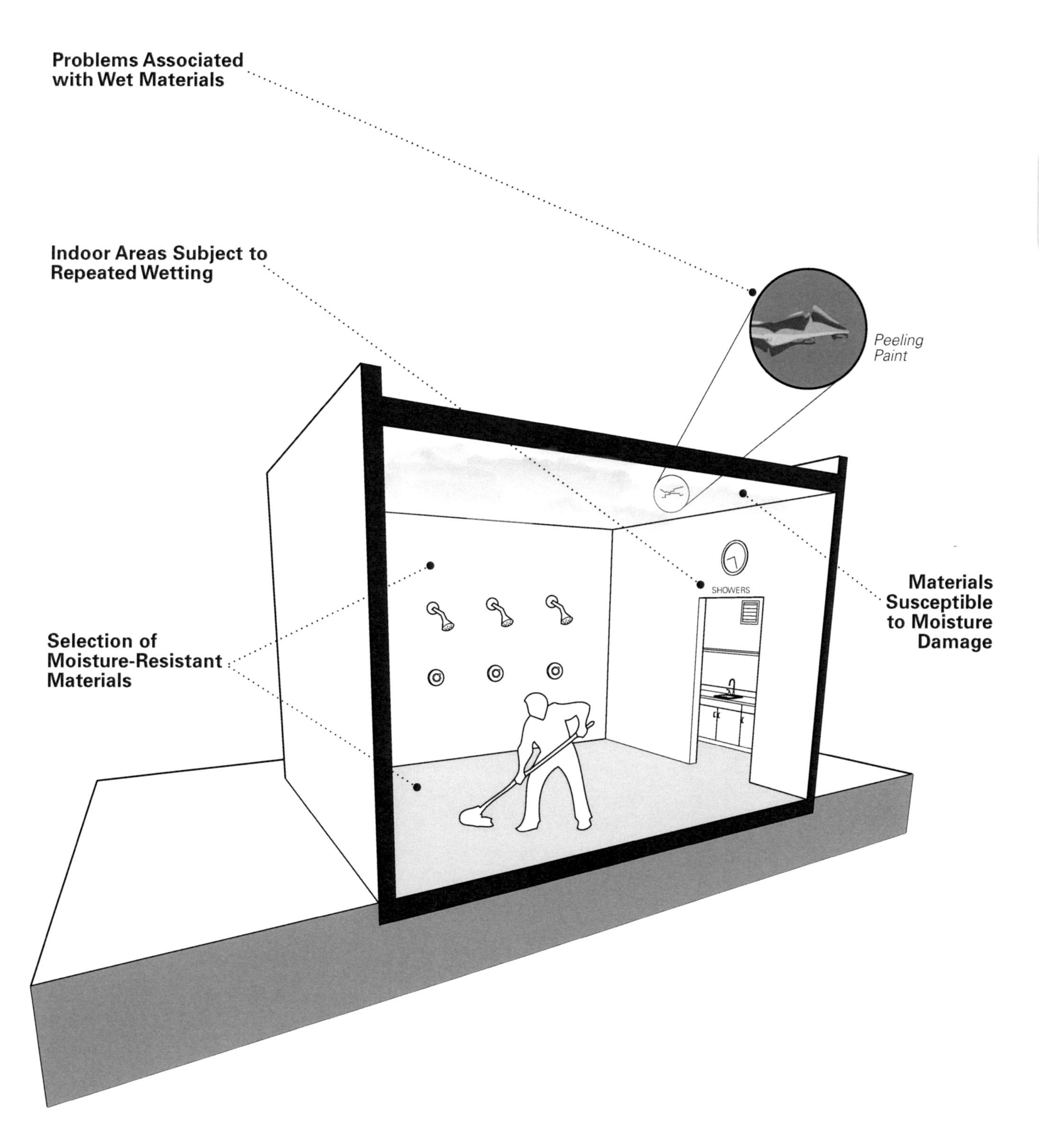
Problems Associated with Wet Materials
Indoor Areas Subject to Repeated Wetting
Peeling Paint
SHOWERS
Materials Susceptible to Moisture Damage
Selection of Moisture-Resistant Materials

## Building Entrance Design as Component in Control of Wet Surfaces and Materials

**Figure 2.5-A** Inadequate Track-Off Design for Local Conditions
*Photograph courtesy of Hal Levin.*

Depending on the local climate and season, building entrance areas may be exposed to wet conditions on a recurring basis. A social services building in a cold winter climate provides an example where inappropriate selection of flooring materials coupled with poor entranceway design created an IAQ problem that was further exacerbated by HVAC design and operation (Figure 2.5-A).

This building's inadequate track-off system could not cope with the snow and moisture introduced by visitors' footwear. Clients for the social services department waited in a large carpeted corridor immediately adjacent to the building entrance. Dirt and moisture accumulated on the corridor's carpeting such that it typically remained wet. A failure in the building's air-handling system resulted in reduced outdoor air delivery during cold winter days. As a result, inadequate air exchange limited the drying potential for the wet carpeting. The combined effect of these factors was an extremely moldy carpet and a strong odor throughout the social services section. Integrated design employing proper track-off systems, use of appropriate flooring materials, and effective HVAC system design and operation would have prevented these IAQ problems.

2.5

# Consider Impacts of Landscaping and Indoor Plants on Moisture and Contaminant Levels



There are potential advantages and disadvantages associated with the presence of plants as a component of the building envelope (e.g., green roofs or roof gardens, living facades, or vertical gardens) or on walls or other locations in the interior space (e.g., atrium gardens, living walls, vertical gardens, biowalls). As part of their physiology, plants emit water molecules into the air through the process of transpiration. In an outdoor environment like a building roof, this provides evaporative cooling. Plants also provide shading to the microenvironment. Inside buildings, an average-sized houseplant emits up to 0.22 lb (100 g) of water per day into the indoor air. An increased amount of water vapor in the air will raise the relative humidity. In a building, this is an advantage during the dry season depending on the source of the water but can be a disadvantage in a warm, humid condition if not well managed.

Benefits of green roofs are thought to include reduction in stress (i.e., thermal stress) on the water proofing membrane, reduction in heat island effects in urban areas, and reduction in storm water runoff. It is widely recognized that both the integrity and protection of the waterproofing membrane beneath the roof garden need to be of very high quality if leakage into the building is to be avoided. In this regard, the building architect and the landscape (garden) architect need to work together to ensure that the waterproofing membrane is installed with excellent workmanship and that penetrations through the membrane are avoided.

The presence of indoor flora (potted plants, atrium gardens, etc.) is generally perceived as beneficial to occupants. However, this assumes that the water transpired does not exceed the capacity of the HVAC system to manage the increased water in the room, that the potted plants are not overwatered, and that atrium gardens are well maintained. Additionally, there is limited research that suggests that root-zone microbial communities of indoor plants reduce VOC contaminants in the indoor air. However, the presence of indoor plants needs to be decided with caution, because some literature also shows that potted plants can result in elevated levels of some fungi indoors, including some pathogenic species.

The illustration in Figure 2.6-A provides background for understanding the concept of water activity ($a_w$), which is a measure of how readily microorganisms or plant roots can extract moisture or free water for growth from the materials on which they are growing. Many fungi and bacteria can grow at an $a_w$ of 0.97–0.98. This is equivalent to the moisture content in a building material in a closed system that has equilibrated with a 97%–98% RH atmosphere in that closed system. The roots of green plants require an $a_w$ of 0.97–0.98 in order to extract water molecules from the materials on which they are growing. These conditions are similar to those that allow for microbial growth on construction materials. Thus, the concept that indoor plants may be beneficial needs to be tempered with the realization that moist building materials and the root-zone are also growth sites of various microbial communities.

## Strategy 2.6

2.6




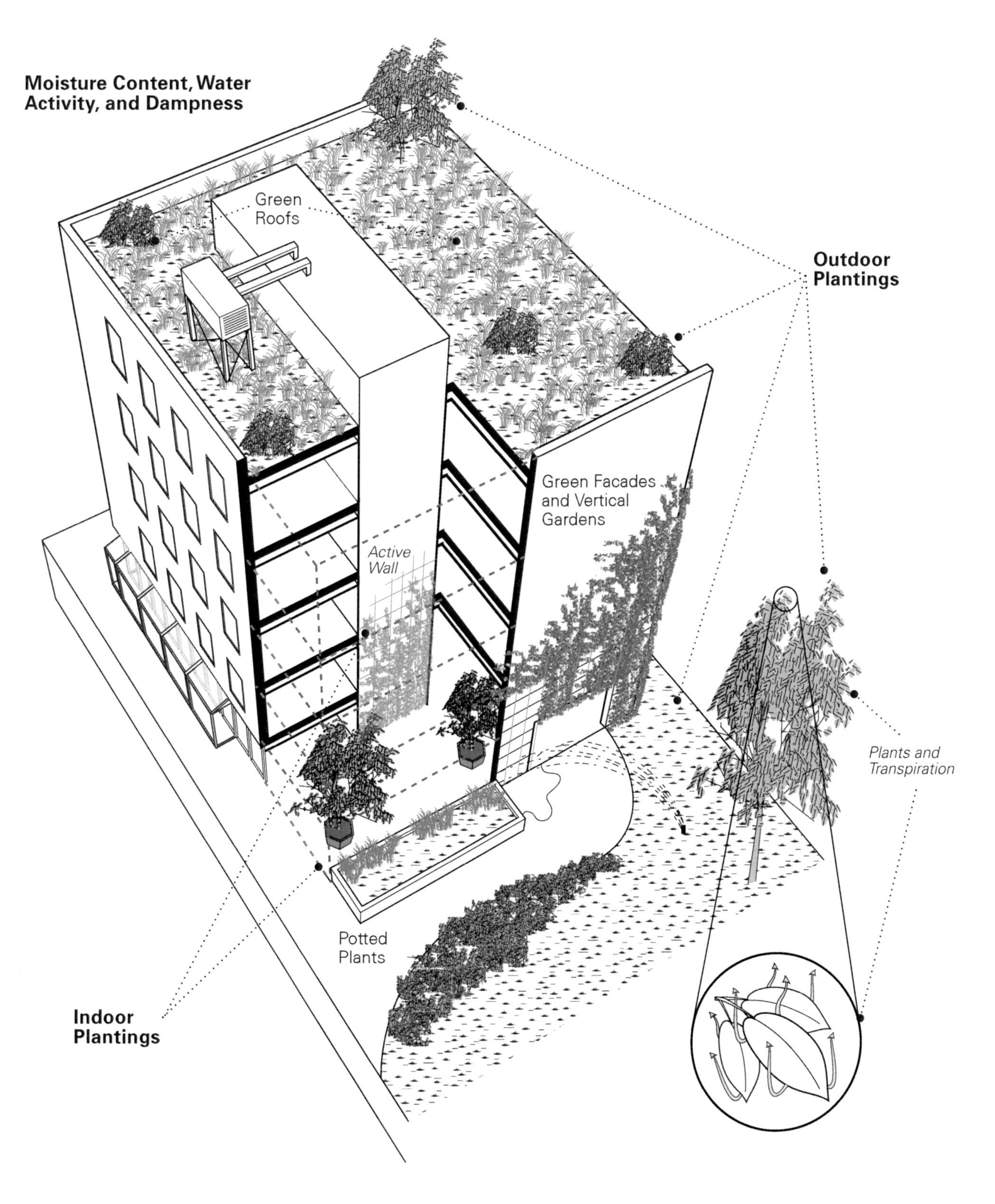

Moisture Content, Water Activity, and Dampness
Green Roofs
Outdoor Plantings
Green Facades and Vertical Gardens
Active Wall
Plants and Transpiration
Potted Plants
Indoor Plantings

## Moisture in Envelope Infrastructure Facilitates Growth of Tree Sapling

**Figure 2.6-A** Tree Sapling Growing from Upper Portion of Window
*Photograph courtesy of Phil Morey.*

Figure 2.6-A shows a tree sapling growing from the upper portion of a window. The roots of this plant are obtaining moisture (see discussion on water activity in the overview) and nutrients from the envelope infrastructure. Microorganisms including fungi and bacteria also grow on plant roots in and on organic construction materials, from which the plant root systems are extracting water.

2.6

# Limit Entry of Outdoor Contaminants

Contaminants from outdoor sources can have a major influence on IAQ. These contaminants include particles and gases in outdoor air, contaminants in the soil and groundwater, herbicides and pesticides applied around the building, and contaminants carried in by pests. The Strategies in this Objective are intended to limit entry of these contaminants.

Outdoor air pollutants entering a building through ventilation and infiltration can have significant health impacts. For example, airborne particles and ozone are both associated with respiratory and cardiovascular problems ranging from aggravation of asthma to premature death in people with heart or lung disease. In many areas of the U.S., levels of these and other pollutants exceed standards set by the EPA. Even in areas where regional outdoor air quality is good, pollution may be high at specific sites due to local sources.

- Strategy 3.1 – Investigate Regional and Local Outdoor Air Quality describes assessment of outdoor air pollution levels and control measures to limit the entry of these contaminants.
- Strategy 3.2 – Locate Outdoor Air Intakes to Minimize Introduction of Contaminants addresses separation of air intakes from such local and on-site sources as motor vehicle exhaust, building exhausts, and cooling towers.
- Strategy 3.3 – Control Entry of Radon describes mitigation techniques for radon, a naturally occurring radioactive soil gas that is the second leading cause of lung cancer in the U.S.
- An important but less widely recognized source of contaminants is intrusion of vapors from contaminated soil or groundwater. Strategy 3.4 – Control Intrusion of Vapors from Subsurface Contaminants describes processes to screen sites for such sources and techniques to limit vapor intrusion.
- People entering buildings can track in contaminants such as pesticides as well as dirt and water that can foster microbial growth. Strategy 3.5 – Provide Effective Track-Off Systems at Entrances describes strategies to reduce tracked-in pollutants.
- Rodents, birds, insects, and other pests can be sources of infectious agents and allergens. Strategy 3.6 – Design and Build to Exclude Pests describes techniques to limit infestation by pests.

Other Strategies that are important in limiting entry of outdoor contaminants are the following:

- Strategy 1.1 – Integrate Design Approach and Solutions
- Strategy 2.3 – Maintain Proper Building Pressurization
- Strategy 2.6 – Consider Impacts of Landscaping and Indoor Plants on Moisture and Contaminant Levels
- Strategy 4.4 – Control Legionella in Water Systems
- Strategy 7.5 – Provide Particle Filtration and Gas-Phase Air Cleaning Consistent with Project IAQ Objectives

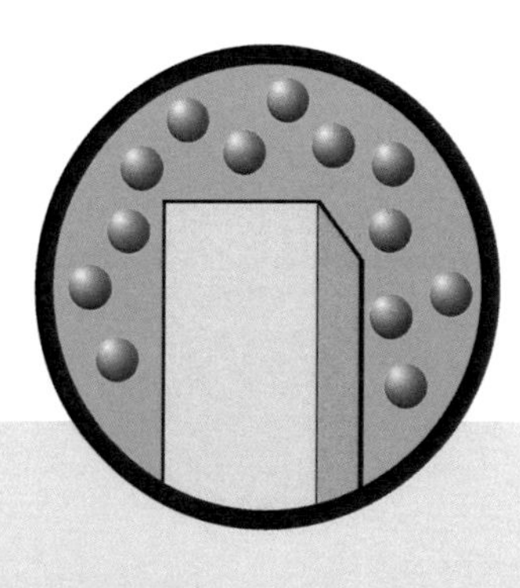

# Objective 3

**3.1** Investigate Regional and Local Outdoor Air Quality

**3.2** Locate Outdoor Air Intakes to Minimize Introduction of Contaminants

**3.3** Control Entry of Radon

**3.4** Control Intrusion of Vapors from Subsurface Contaminants

**3.5** Provide Effective Track-Off Systems at Entrances

**3.6** Design and Build to Exclude Pests

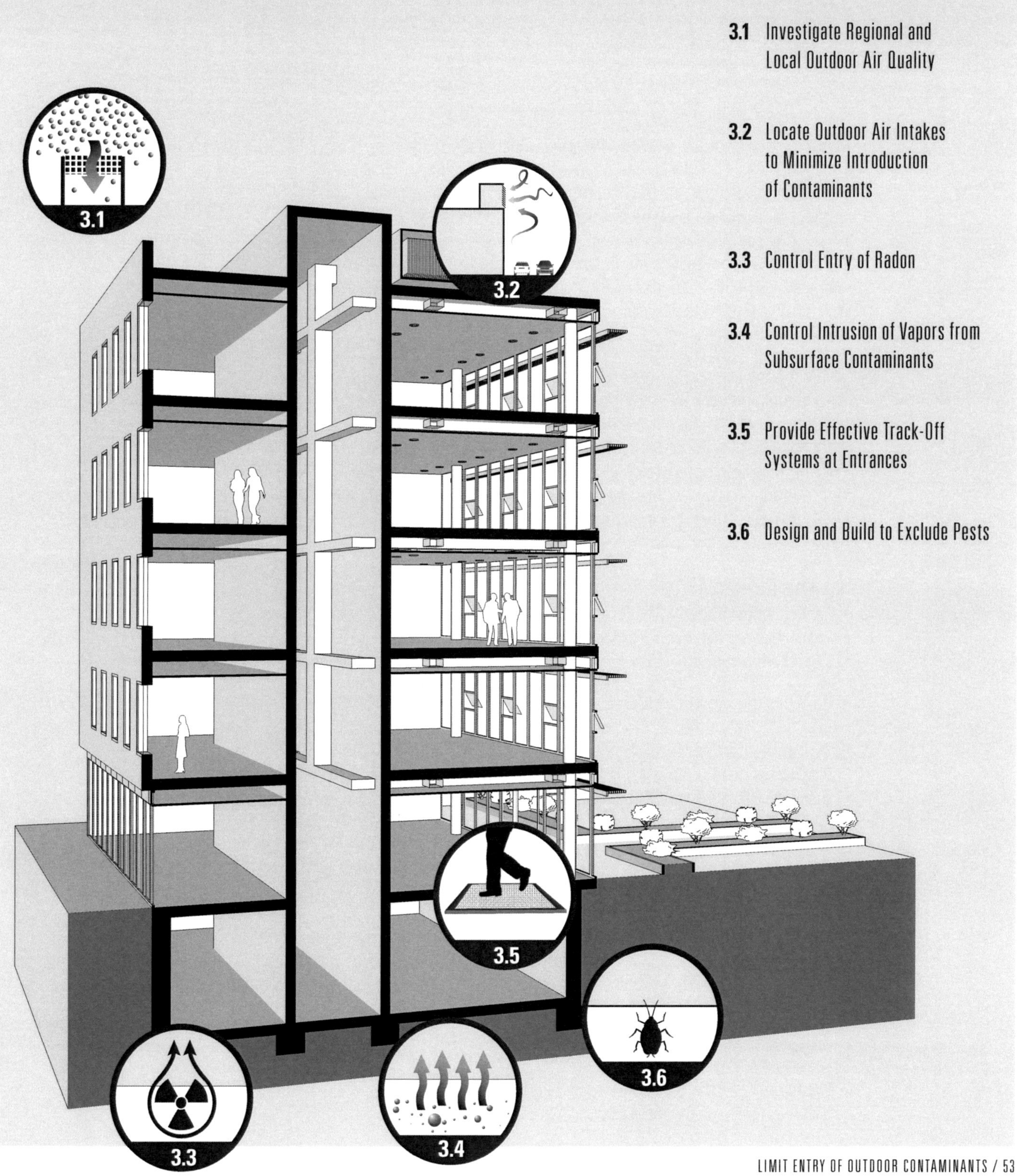

# Investigate Regional and Local Outdoor Air Quality

**Introduction**
**Assessment**
- Determine Compliance with NAAQS
- Determine Whether Local Sources are Present

**NAAQS Particles**
- Particulate Matter — PM10
- Particulate Matter — PM2.5
- Lead

**NAAQS Gases**
- Ozone
- Nitrogen Dioxide ($NO_2$)
- Sulfur Dioxide ($SO_2$)
- Carbon Monoxide (CO)

**Other Pollutants**
- Dust
- Volatile Organic Compounds (VOCs)
- Odors

**References**

## Assessment

> *"The control of air quality is one of the functions of complete air conditioning, and some knowledge of the composition, concentration and properties of air contaminants under various circumstances is therefore essential... the engineer will find, at times, that odors originating outside buildings in industrial of business districts may have an even greater bearing than indoor contamination on the kind and capacity of equipment he must provide for a high quality air supply installation." (ASHVE 1946)*

More than sixty years after the 1946 reference above, the outdoor atmosphere still contains many particles and gases that can adversely affect IAQ. A primary resource for information on outdoor air pollution is in the Green Book on the EPA Web site (www.epa.gov/air/oaqps/greenbk). EPA illustrates on maps areas that are not in compliance (nonattainment) with the National Ambient Air Quality Standards (NAAQS) (EPA 2008b, 2008c). EPA established the NAAQS as directed by Congress in the Clean Air Act. Pollution can be from particles, gases, or both.

Following the requirements in ASHRAE Standard 62.1 (ASHRAE 2007a), the first step in ventilation design for IAQ is to determine compliance with outdoor air quality standards in the region where the building will be located. The next step is to determine if there are any local sources of outdoor air pollution that may affect the building. Filtration or air cleaning can then be considered as a means of reducing the entry of these outdoor contaminants into the indoor environment. Operating scenarios can also be developed to reduce entry of pollutants into the building if the pollutant levels vary over time. For instance, CO from cars will vary with traffic volumes and patterns. Ozone also varies by time of day, with higher concentrations usually in the afternoon.

### NAAQS Particles

- Particles designated PM10 are particles that are smaller than 10 µm in diameter. ASHRAE Standard 62.1-2007 requires a minimum of MERV 6 filters at the outdoor air in areas that are nonattainment with PM10. Higher Minimum Efficiency Reporting Value (MERV) filters will provide additional filtration efficiency.

- Particles designated PM2.5 are particles that are smaller than 2.5 µm in diameter. Filters tested by *ANSI/ASHRAE Standard 52.2, Method of Testing General Ventilation Air-Cleaning Devices for Removal Efficiency by Particle Size* (ASHRAE 2007c), are measured for efficiency at particle size fractions including 0.3 to 10 µm. Filters need to have MERV values greater than MERV 8 to have any effective removal efficiency on these smaller particles. Filters with MERV ≥ 11 are much more effective at reducing PM2.5.

- Lead is a solid and will be a particle or may be attached to other particles in the atmosphere. Filters that are effective on small particles will also be effective at removing lead from the outdoor airstream.

### NAAQS Gases

- Ozone is formed in the atmosphere by a photochemical reaction under sunlight. Therefore, ozone is not generated on cloudy or cold days. Ozone air treatment is provided by carbon or other sorbent filters that cause the ozone to react on the surface. Table 3.1-A illustrates the air quality index for ozone.

- There are no areas in the U.S. that are currently nonattainment for nitrogen dioxide ($NO_2$). There are gas-phase air cleaners that can be effective on $NO_2$.

- Sulfur dioxide ($SO_2$) can be cleaned by gas-phase air cleaners. Certain filter materials (for example, activated alumina/$KMnO_4$) adsorb $SO_2$.

# Strategy 3.1

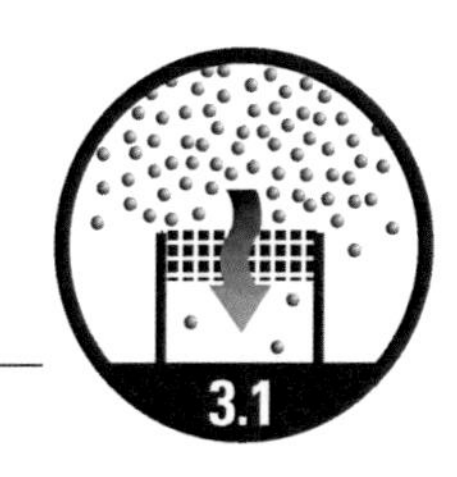
3.1

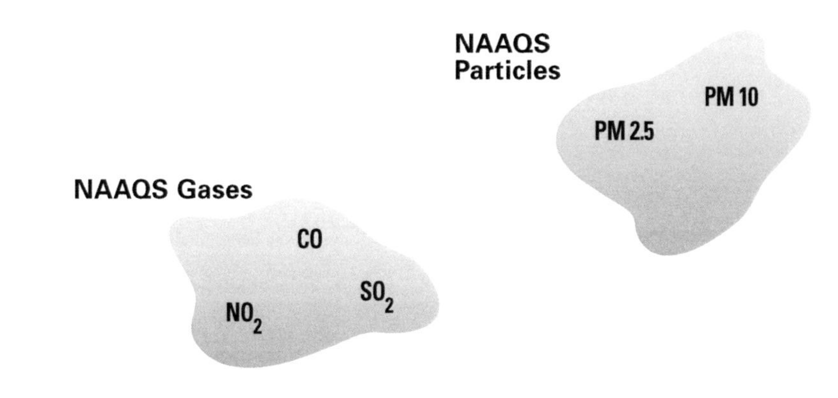
NAAQS Particles
PM 10
PM 2.5
NAAQS Gases
CO
$SO_2$
$NO_2$

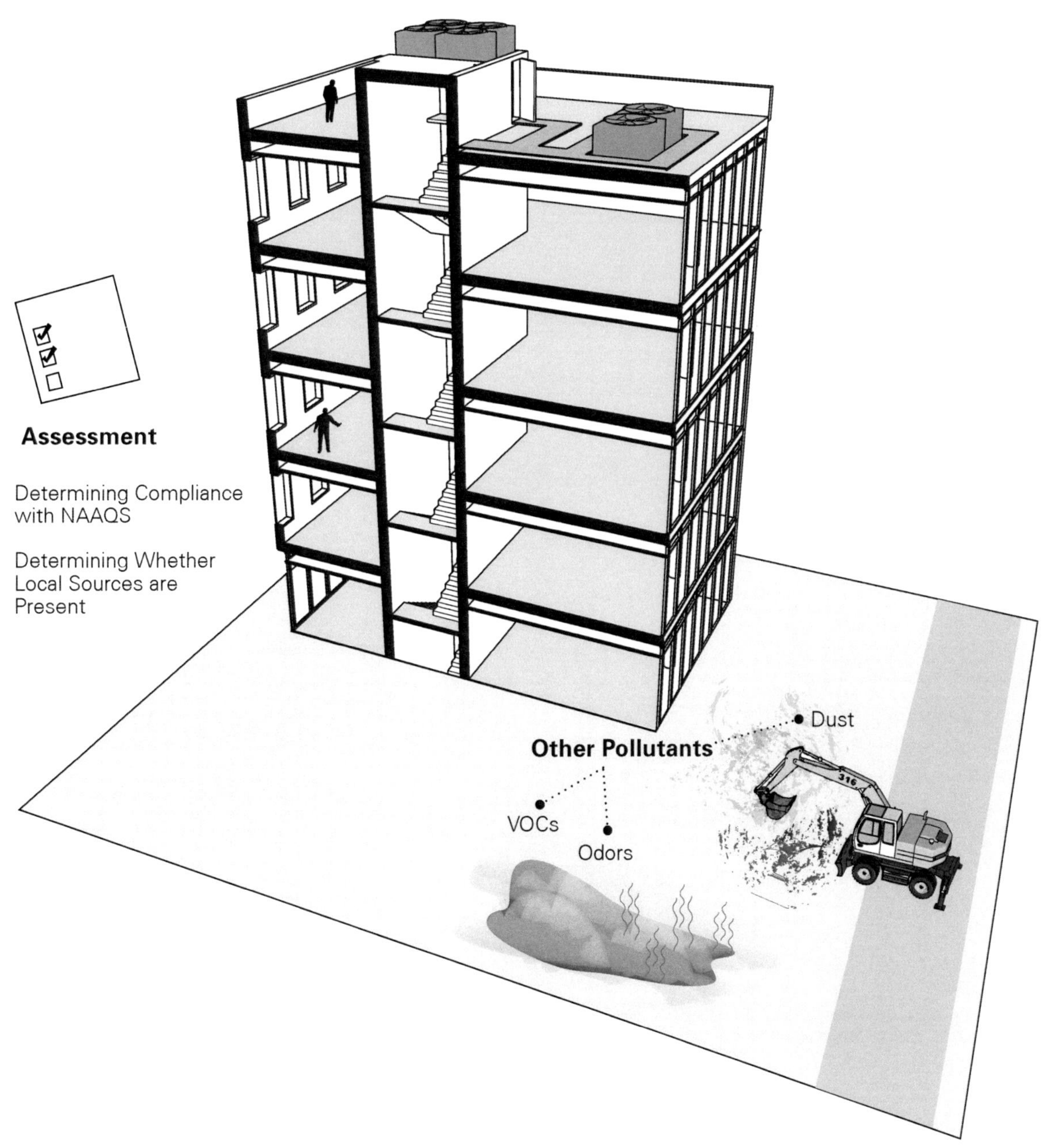
Assessment
Determining Compliance with NAAQS
Determining Whether Local Sources are Present
Other Pollutants
Dust
VOCs
Odors
316

- There is no commercially available air cleaner for CO that operates at room temperature. Scheduling of activities and the ventilation system operation, as well as outdoor air intake location, are strategies to reduce the impact of CO on the indoor environment.

**Other Pollutants**

- Airborne dust is no longer regulated as a NAAQS pollutant but can be a problem in areas with agriculture, high pollen, certain industries, or desert climates. Filtration of airborne dust needs to focus on the dust-holding capacity of the filtration system.

- Outdoor sources of VOCs include industrial emissions, traffic, mobile equipment, area sources such as wastewater lagoons, and some natural sources. If there are local (nearby) sources of VOCs, filtration or air cleaning needs to be considered.

- Odors in the atmosphere are often (but not always) regulated in response to citizen complaints in urban environments. Odors can be removed from outdoor air with air-cleaning technology that is tailored to the specific compounds that cause the odor. Occupants tend to be highly sensitive to odors.

## Controlling Outdoor Air Pollutants Indoors

**Figure 3.1-A** Building in a Polluted City
*Photograph courtesy of H.E. Burroughs.*



Outdoor ozone pollution results in several adverse effects, some of which are lung irritation and respiratory illness. Ozone can also damage paper documents and books, which is of great concern when they are valuable. Ozone and acid gases (e.g., gases from sulfur) are detrimental to the chemistry of paper, and prolonged exposure to trace concentrations can cause fading and embrittlement. When the state archive facility shown in Figure 3.1-A was designed, special consideration was given to controlling outdoor air pollutants. In separate filtration systems, both the outdoor air and the recirculated air are treated with deep-bed gas-phase air-cleaning equipment as well as high-efficiency MERV 16 particulate filters. MERV 6 pleated particulate filters are used to prefilter the final filters. A special dehumidification system is also employed to remove the excess humidity from the outdoor air. The archives building is located near major expressways and is beneath a primary landing pathway for the Atlanta international airport. When evaluated in 2007, the ozone concentration of the outdoor air was tested at peaks of 88 ppb (172 $\mu g/m^3$), which is sufficient to cause deterioration of paper. Yet concentrations of ozone in the supply to the storage chambers were below detection. Further, there were no sulfur compounds found in the conditioned space.

| Air Quality Index | Protect Your Health |
|---|---|
| Good (0-50) | No health impacts are expected when air quality is in this range. |
| Moderate (51-100) | Unusually sensitive people should consider limiting prolonged outdoor exertion. |
| Unhealthy for Sensitive Groups (101-150) | The following groups should limit prolonged outdoor exertion:<br>• People with lung disease, such as asthma<br>• Children and older adults<br>• People who are active outdoors |
| Unhealthy (151-200) | The following groups should avoid prolonged outdoor exertion:<br>• People with lung disease, such as asthma<br>• Children and older adults<br>• People who are active outdoors<br>Everyone else should limit prolonged outdoor exertion. |
| Very Unhealthy (201-300) | The following groups should avoid all outdoor exertion:<br>• People with lung disease, such as asthma<br>• Children and older adults<br>• People who are active outdoors<br>Everyone else should limit outdoor exertion. |

**Table 3.1-A** Air Quality Index for Ozone
*Source: OAQPS (2009).*

# Locate Outdoor Air Intakes to Minimize Introduction of Contaminants



Outdoor air enters a building through its air intakes. In mechanically ventilated buildings, the air intakes are part of the HVAC system. In naturally ventilated buildings, the air intakes can be operable windows or other openings in the building's envelope.

As outdoor air enters a building through its air intakes, it brings with it any contaminants that exist outside the building near the intake. That is why the quality of the outdoor air delivered to a building greatly affects the quality of the indoor air. Therefore, it is important to evaluate the ambient air quality in the area where a building is located as well as the presence of local contaminant sources. Outdoor air intakes need to be designed and located in such a way as to reduce the entrainment of airborne pollutants emitted by these sources.

**Applicable Codes, Standards, and Other Guidance**

Mechanical codes—such as *International Mechanical Code* (*IMC*; ICC 2006a), *International Plumbing Code* (*IPC*; ICC 2006b), *Uniform Mechanical Code* (*UMC*; IAPMO 2006a), and *Uniform Plumbing Code* (*UPC*; IAPMO 2006b)—have some requirements for the locations of building intakes. However, in most cases these requirements are very limited and there may be value in considering going beyond these requirements.

Table 5.1 of ASHRAE Standard 62.1 (ASHRAE 2007a) lists minimum separation distances between air intakes and specific contamination sources. Although ASHRAE Standard 62.1 does not cover all possible sources, it does give the designer a guiding tool. Appendix F of the same standard allows the designer to calculate distances from sources other than the ones listed in Table 5.1. The distances listed in ASHRAE Standard 62.1 should be considered design minimums; greater distances may provide better protection against these contaminants entering the building.

**Cooling Towers, Evaporative Condensers, and Fluid Coolers**

According to Table 5.1 of ASHRAE Standard 62.1, outdoor air intakes need to be located at least 25 ft (7.6 m) from plume discharges and upwind (prevailing wind) of cooling towers, evaporative condensers, and fluid coolers. In addition, outdoor air intakes need to be located at least 15 ft (4.6 m) away from intakes or basins of cooling towers, evaporative condensers, and fluid coolers. Buildings designed with smaller separation distances can increase the risk of occupant exposure to *Legionella* and other contaminants, such as the chemicals used to treat the cooling tower water. See Strategy 4.4 - Control *Legionella* in Water Systems for more information.

**Other Sources of Contamination**

All nearby potential odor or contaminant sources (such as restaurant exhausts, emergency generators, etc.) and prevailing wind conditions need to be evaluated. Locations of plumbing vents in relationship to outdoor air intakes in high-rise buildings may require additional analysis. Model codes such as *IMC* and *UMC* require a 3–10 ft (0.9–3.0 m) separation distance between building air intakes and terminations of vents carrying non-explosive or flammable vapors, fumes, or dusts. In the case of plumbing vents, *IPC* and *UPC* require a 2–10 ft (0.6–3.0 m) separation distance. For the health-care industry, *Guidelines for Design and Construction of Health Care Facilities* (AIA 2006) requires separation distances of 25 ft (7.6 m) between building intakes and plumbing vents, exhaust outlets of ventilating systems, combustion equipment stacks, and areas that may collect vehicular exhaust or other noxious fumes. However, these guidelines allow the 25 ft (7.6 m) separation distance to be reduced to 10 ft (3 m) if plumbing vents are terminated at a level above the top of the air intake.

# Strategy 3.2

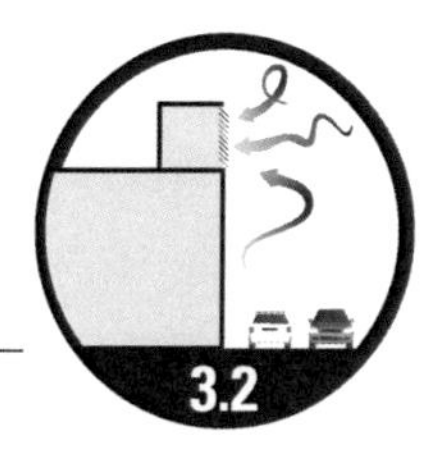


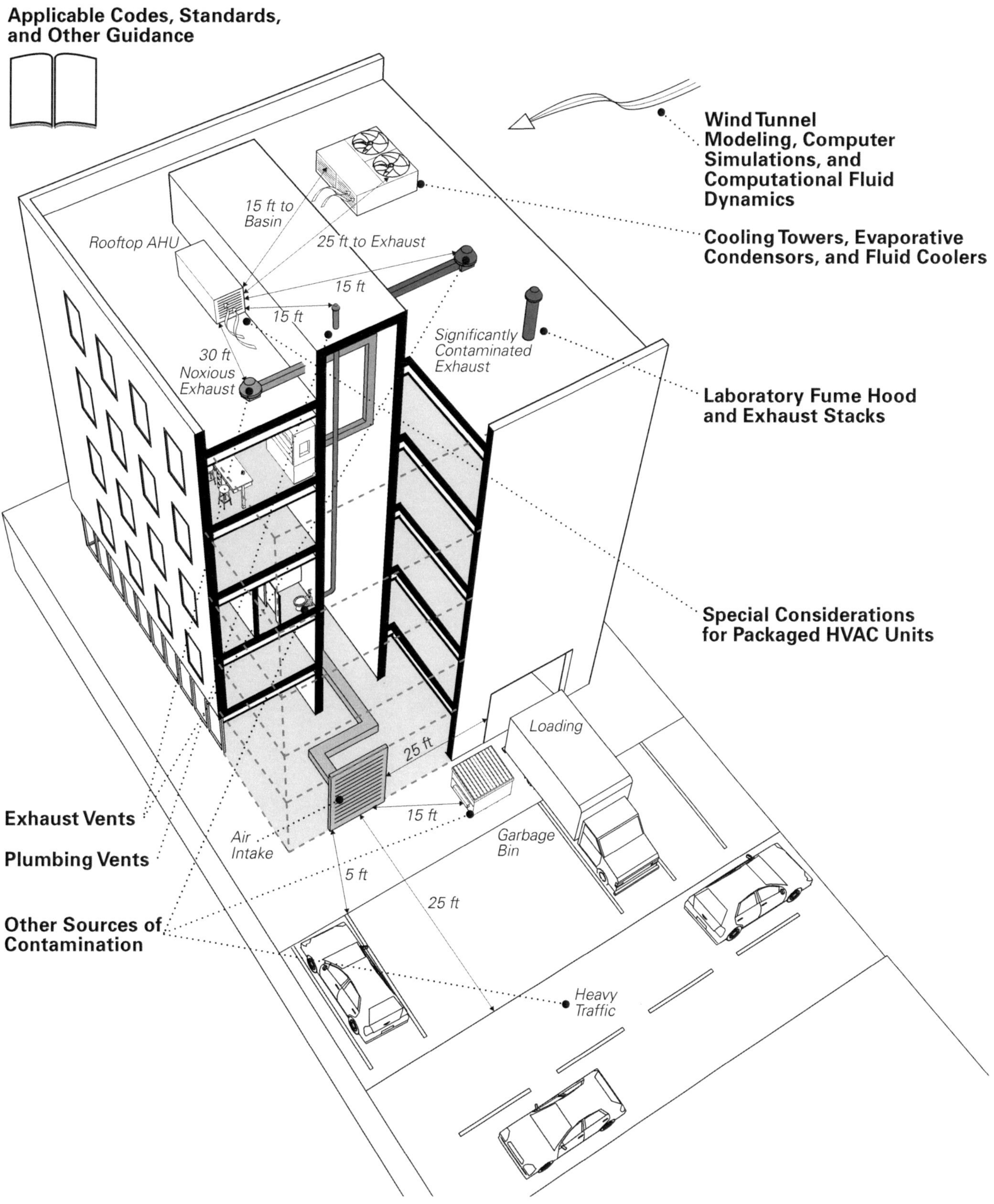


STRATEGY OBJECTIVE 3.2

### Modeling

It is clear that due to wind effects around buildings and multiple other local variables, establishing separation distances that will result in no entrainment for each source is extremely difficult if not impossible. Each design case must be evaluated based on local conditions and variables, and the designer ultimately needs to exercise professional judgment. In some cases, advanced calculations and/or modeling may be required, such as wind tunnel analyses with scale models, computer simulations, or computational fluid dynamics (CFD) analyses.

### Packaged HVAC Systems

In packaged HVAC systems, an exhaust stack elevated 10 in. (0.25 m) or more can reduce re-entrainment of combustion products. In HVAC systems where the intakes and exhausts are in close proximity, dilution of building exhaust air in the economizer mode is significantly less than the dilution of flue gas in the heating mode. Packaged HVAC units need to be located so that their air intakes and exhausts are directed away from large obstructions.

## Intake/Exhaust Separation at a New Office Building

The HVAC unit shown in Figure 3.2-A is one of several units on the roof of a large five-story office building. Indirect evaporative cooling was installed in all HVAC units before the building was occupied. Although the reason for the relocation of all the intakes was the installation of the indirect evaporative coolers, it created an opportunity for the designers to increase the separation distance between intakes and exhausts.

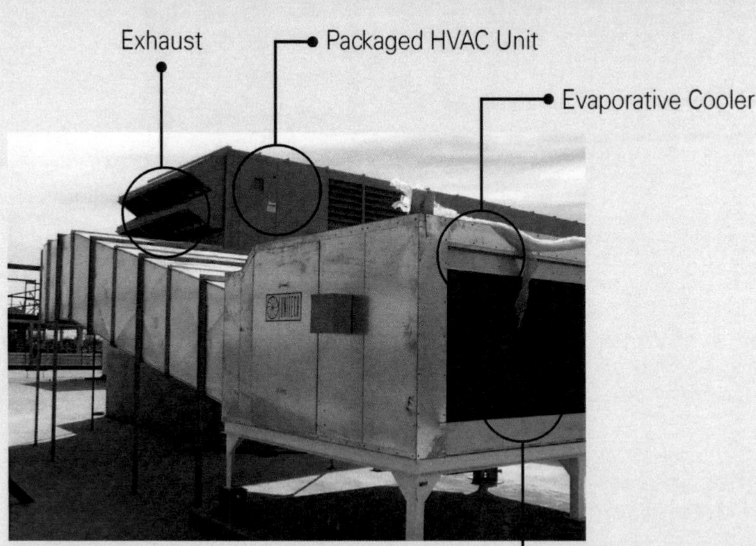


**Figure 3.2-A** Large Packaged HVAC Unit with Relocated Outdoor Air Intake to Reduce Exhaust Air Re-entrainment (Shown is a Closeup of the Indirect Evaporative Cooler at the Outdoor Air Intake)
*Photograph courtesy of Leon Alevantis.*

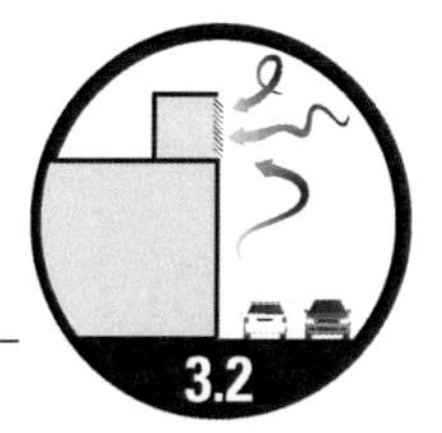
3.2

# Control Entry of Radon



Radon is a radioactive gas formed from the decay of uranium in rock, soil, and groundwater. Exposure to radon is the second leading cause of lung cancer in the U.S. after cigarette smoking and is responsible for about 10% to 14% of lung cancer deaths (NAS 1999).

Radon most commonly enters buildings in soil gas that is drawn in through joints, cracks, or penetrations or through pores in concrete masonry units (CMUs) when the building is at negative pressure relative to the ground. The potential for high radon levels varies regionally, with additional variation from building to building in the same region and even from room to room in the same building.

Design for control of radon entry includes three components, as follows:

- Active soil depressurization (ASD), which uses one or more suction fans to draw radon from the area below the building slab and discharge it where it can be harmlessly diluted to background levels. By keeping the sub-slab area at a lower pressure than the building, the ASD system greatly reduces the amount of radon-bearing soil gas entering the building. A permeable sub-slab layer (e.g., aggregate) allows the negative pressure field created by a given radon fan to extend over a greater sub-slab area.
- Sealing of radon entry routes, including ground-contact joints, cracks, and penetrations and below-grade CMU walls.
- Use of HVAC systems to maintain positive building pressure in ground-contact rooms and to provide dilution ventilation, as an adjunct to ASD and sealing of radon entry routes.

An ASD system and sealing of radon entry routes are easier and cheaper to implement during construction than they are as a retrofit. In addition, many elements of a radon control system are also useful for reducing the intrusion of vapors from brownfield sites and for reducing the penetration of liquid water and water vapor into below-grade building assemblies. For these reasons, it may be preferable to address radon during initial design and construction rather than after the building is built.

The design team needs to review with the owner the radon potential, the synergies of radon control techniques with design for other IAQ issues, and the new construction vs. retrofit costs to determine whether to implement radon control measures.

# Strategy 3.3

3.3

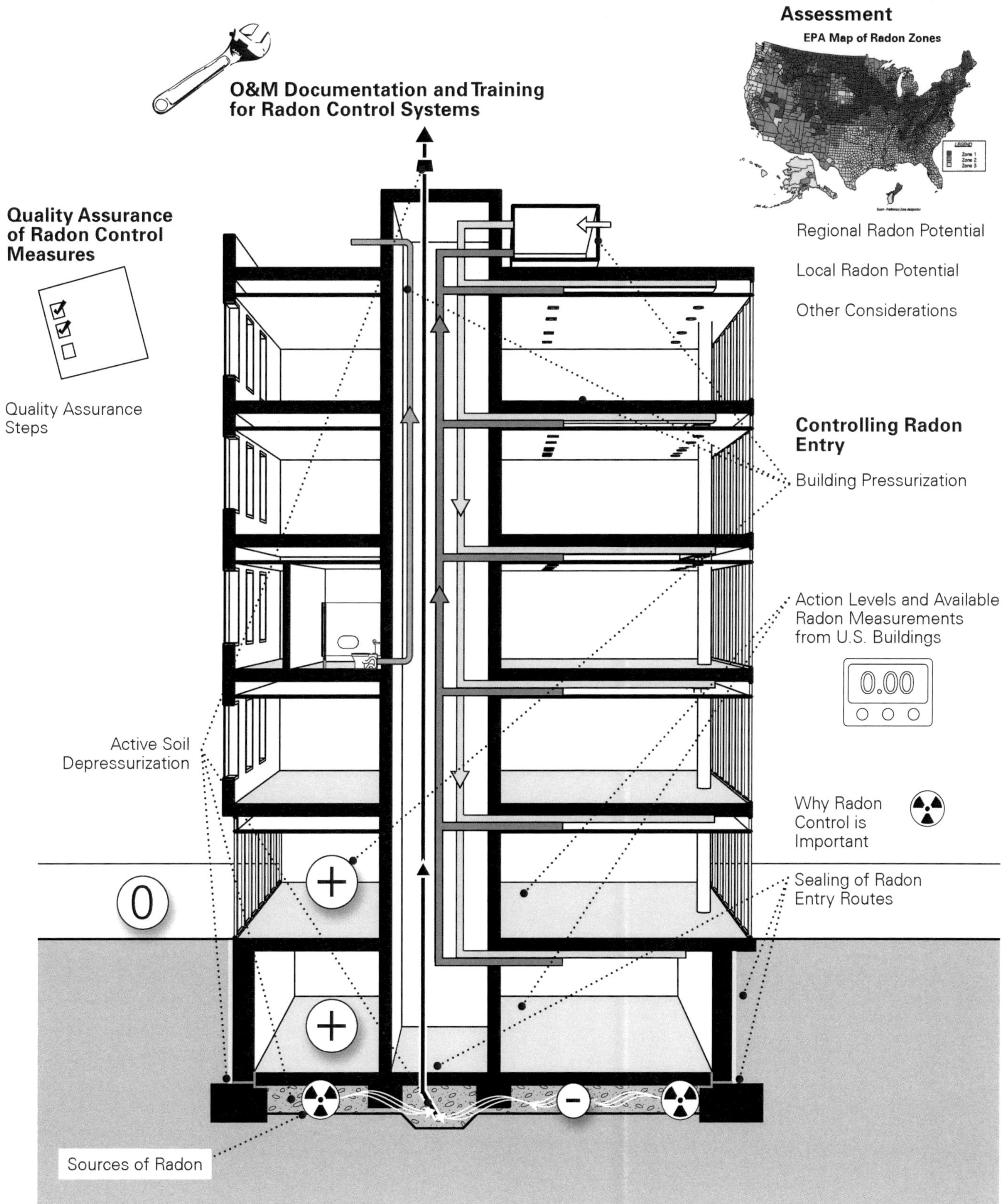
O&M Documentation and Training for Radon Control Systems
Assessment
EPA Map of Radon Zones
Regional Radon Potential
Local Radon Potential
Other Considerations
Quality Assurance of Radon Control Measures
Quality Assurance Steps
Controlling Radon Entry
Building Pressurization
Action Levels and Available Radon Measurements from U.S. Buildings
0.00
Active Soil Depressurization
Why Radon Control is Important
Sealing of Radon Entry Routes
0
Sources of Radon

## How Complaints of Headache and Nausea led to Discovery of High Radon Levels

**Figure 3.3-A** Office Building with Elevated Radon Levels, Identified During an Investigation of Occupant Complaints Related to Another Contaminant
*Photograph courtesy of H.E. Burroughs.*

A four-story owner-occupied office building in the Southeastern U.S. (Figure 3.3-A) is located on a site with shallow topsoil over granite in an identified radon region. At the time of the investigation, the building was eight years old. It has an open atrium lobby that penetrates all floors. Each floor is served by a separate air handler, and ventilation and makeup air are introduced from the roof through a common shaft to each mechanical room. The return air on each floor flows to the lobby atrium, entering each floor's mechanical room through transfer grilles from the lobby wall. The lower floors are open-plan office space, and the upper floor is devoted to the company cafeteria and the executive dining room. The lobby is decorated with architectural plantings including a large and valuable ficus tree several stories tall.

An investigation was undertaken in response to complaints of regular, repeated occurrences of headaches and nausea reported by over 50% of the occupant population every other Friday afternoon after 2:00 p.m. Air testing conducted on a "problem" Friday revealed elevated VOC levels (above 5000 μg/m$^3$) and high fungal counts (greater than three times the outdoor levels). To the surprise of the investigators, radon levels were also elevated to over three times the level at which EPA recommends action be taken (EPA 2009). When the air testing was done, the investigation team reviewed the airflow data from specifications and the test and balance report but took no airflow measurements.

The report and findings were submitted by the investigation team two weeks later, on another "problem" Friday afternoon. At this time, it was obvious that fish was being served in the cafeteria because the distinctive odor was present on all floors of the building, clearly migrating from the fourth-floor kitchen. Based on this obvious evidence of contamination crossover between floors, the kitchen area and operating practices were investigated. It was found that the kitchen closed at 2:00 p.m. on Friday afternoons, and on every other Friday pesticide treatment was applied. The pesticide aerosol quickly migrated throughout the building because of the lobby atrium's negative pressurization and was drawn onto all floors in the return air. The negative atrium pressure was also depressurizing the soil in which the ficus tree was planted. The team learned that the root system penetrated the concrete slab because of the size of the tree. Thus, there was no air barrier between the soil and the space, which allowed the depressurization to draw radon into the building. The soil was also the source of the high fungal counts.

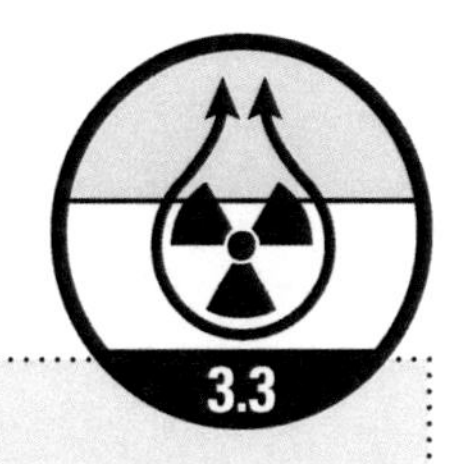


Once the causes of these problems were determined, they were easily solved, through:

- the addition of localized exhaust in the kitchen with integral makeup air,
- rescheduling of pesticide application to off-hours,
- provision of additional fan-powered outdoor air delivered directly to each floor's mechanical room, and
- balancing of outdoor air quantities to deliver a larger quantity to the ground-floor zone.

The atrium lobby pressure relative to the outdoors was about −0.2 to −0.24 in. w.g. (−50 to −60 Pa) before the modifications and increased to 0 to 0.032 in. w.g. (0 to 8 Pa) after the modifications. This was sufficient to bring the level of radon right at the source (the ficus tree root area) to below the action level. The problem of headaches and nausea related to the pesticide application was resolved as well. Had the building not had an IAQ problem that was causing acute symptoms in the building occupants, the elevated radon levels likely would never have been detected or resolved.

# Control Intrusion of Vapors from Subsurface Contaminants



### What is Vapor Intrusion?

Vapor intrusion is "the migration of a [chemical of concern] vapor from a subsurface soil or groundwater source into the indoor environment of an existing or planned structure" (ASTM 2008, Section 3.2.53). Although vapors can intrude into buildings from naturally occurring subsurface gases, most guidance, standards, and regulations are specific to vapors from environmental contaminants. Contaminants can be present below ground due to accidental spills, improper disposal, leaking landfills, or leaking underground or aboveground storage tanks (Tillman and Weaver 2005). Some of the most widespread present and past land uses associated with chemicals of concern include gas stations, dry cleaners, former industrial sites that used vapor degreasers or other parts-cleaning chemicals, and manufactured gas plants (ASTM 2008). There are hundreds of thousands of contaminated or potentially contaminated sites in the U.S. (EPA 2002). Vapor intrusion can occur even though a site has undergone or is undergoing remediation. Moreover, the potential for vapor intrusion is not limited to the sites where contaminants were originally released, since contaminants reaching the water table can travel at least several miles in contaminated groundwater plumes, and contaminant vapor can travel shorter distances through the vadose zone above the water table.

### Why Is Vapor Intrusion a Concern?

Vapor intrusion is a concern primarily because of the potential for chronic health effects from long-term exposure to low contaminant concentrations, although in extreme cases vapor concentrations can be high enough to cause acute health effects or explosion hazards (EPA 2002). The chemicals of concern for vapor intrusion are those that are sufficiently volatile to migrate through the soil as a vapor and sufficiently toxic that they may adversely affect human health (EPA 2002). These include many VOCs, some semi-volatile organic compounds (SVOCs), and some inorganic substances such as mercury.

### Standards, Guidance, and Regulations

The potential for intrusion of vapors from these sites into buildings was not widely recognized by U.S. regulators until the 1990s (IRTC 2007), and both the science and the regulatory environment are still evolving. ASTM International (ASTM) recently developed a national standard for assessment of vapor intrusion on properties involved in real estate transactions (ASTM 2008). The standard uses a tiered approach to allow properties with a low risk of vapor intrusion to be screened out quickly and at relatively low cost. The Interstate Technology & Regulatory Council (ITRC) published guidance on both assessment and mitigation in 2007.

Screening and assessment of the potential for vapor intrusion must follow the regulations, policies, and guidance of the relevant jurisdiction and must be performed by a person meeting the qualifications required by that jurisdiction. In many cases this will be an environmental professional outside the primary design team. Obtaining guidance from the appropriate regulatory agency early in the process is the best way to avoid problems and ensure a successful project.

# Strategy 3.4

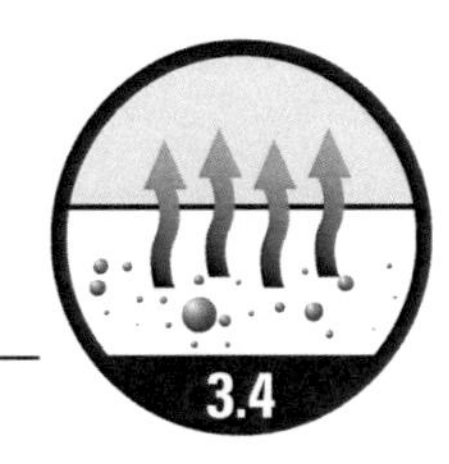


**Screening and Assessment**

Federal Guidance

State Guidance

ASTM E2600-08: A National Standard for Assessment of Vapor Intrusion in Real Estate Transactions

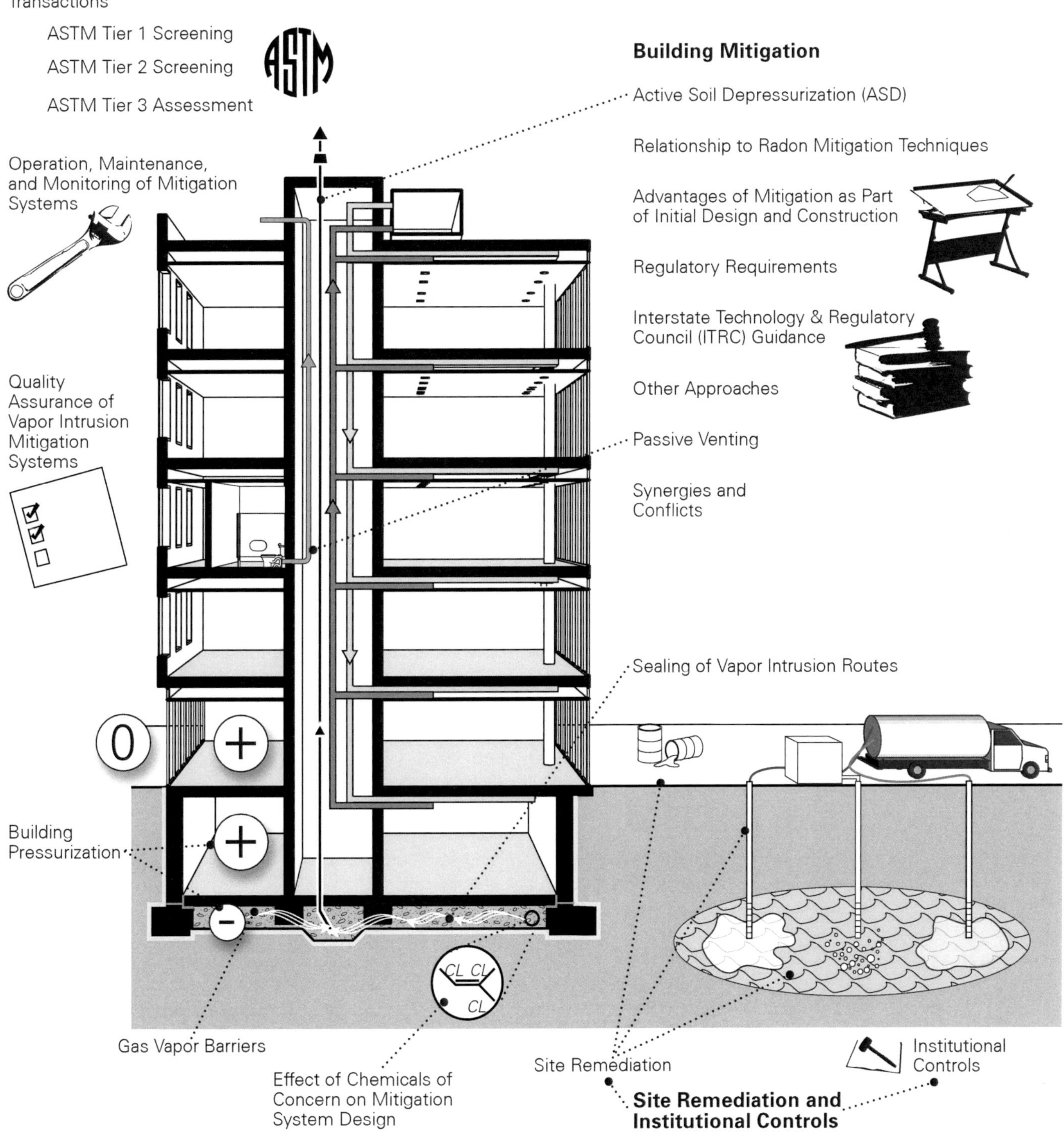

### Site Remediation

In the long term, the best remedy for contaminated sites is to remove the contaminant source or to perform treatments in place to reduce contaminant levels. Some remediation techniques may reduce or eliminate soil gas migration substantially in a relatively short time, rendering building mitigation measures unnecessary. Some other technologies are longer-term strategies that may take many years to reduce the contaminant source to a level where vapor intrusion is no longer a concern. In still other cases, cleanup may be on hold for various reasons. Where longer-term cleanup strategies are used or cleanup is on hold, institutional controls and/or building mitigation are necessary as interim strategies.

### Institutional Controls

Institutional controls are "legally enforceable conditions placed on a property to reduce the likelihood of exposure to unacceptable levels of contaminants" (ASTM 2008, Section 11.3.1). They may include such measures as zoning restrictions, requirements that vapor intrusion mitigation systems be preemptively installed in new construction, requirements that source remediation systems or building mitigation systems be periodically inspected, or requirements that contaminant levels be periodically monitored (ITRC 2007).

### Building Mitigation Systems

Building mitigation systems are systems that reduce intrusion of vapors into buildings.

- *Qualifications for Design and Installation.* These systems must be designed and installed by parties who meet the qualifications of the regulatory agency having jurisdiction and who have relevant expertise. Very few design teams or contractors have this expertise in-house, so most rely on specialized environmental consultants and mitigation contractors. Requirements for design, installation, and performance testing of vapor intrusion mitigation systems vary widely by jurisdiction, so it is important to involve the appropriate regulatory agency early in the design process.

- *When to Address Vapor Intrusion.* Where the need for mitigation is a possibility, it is cheaper and more effective to incorporate the system in the original design and construction rather than to retrofit it after the building is built and vapor concentrations are tested. The need for vapor intrusion control needs to be considered early in design, when other design elements may be modified to potentially eliminate the need or reduce the cost.

- *Comparison with Radon Control.* Vapor intrusion mitigation technologies are largely adapted from radon control strategies. Bringing indoor vapor concentrations below action levels commonly requires reductions of two or three orders of magnitude (and sometimes more), in contrast to the one to two orders of magnitude typically required to bring radon concentrations below action levels. To achieve these larger reductions, it may be necessary to combine several mitigation techniques or to design and install systems to exacting standards.

- *Active Soil Depressurization (ASD).* ASD (sub-slab or sub-membrane) is generally the most reliable and most frequently employed vapor intrusion control technology. ASD systems use one or more suction fans to depressurize the soil in contact with the building. This ensures that the predominant direction of air leakage across those portions of the building envelope in contact with the ground is from the building into the soil rather than from the soil into the building. Soil gas is drawn through vent risers by the suction fan(s) and released above the building, where it can be diluted by ambient air. ASD systems need to be combined with sealing of joints, cracks, and penetrations and sealing of below-grade walls. In new construction, ASD is usually combined with two additional elements to enhance its effectiveness: 1) a "venting layer" of gas-permeable aggregate and/or a network of perforated pipe and 2) a gas vapor barrier above the venting layer. Where the amount of vapor intrusion that will occur is uncertain, it may make sense to install a passive venting system with a gas vapor barrier and add the fan(s) to convert to ASD if post-construction testing shows it to be necessary. Passive systems should always be designed to be readily converted to active systems. In general, use of gas vapor barriers alone without passive venting is not recommended.

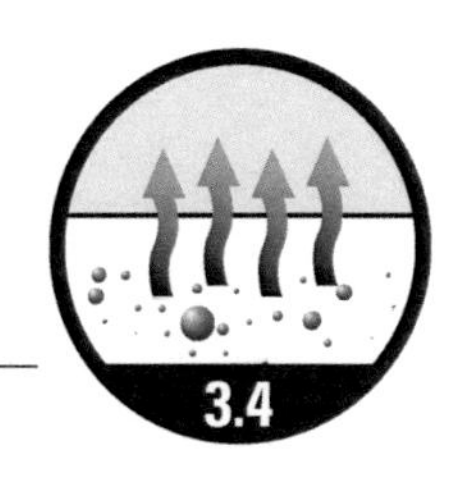


- *Building Pressurization.* Building pressurization can be the best choice for sites with wet or low-permeability soils where ASD is not effective. It can also be a viable alternative to ASD for other buildings if the HVAC system has the capability to reliably maintain positive pressures in all ground-contact rooms. However, pressurization for vapor intrusion control requires a level of quality assurance in design, construction, operation, and maintenance that goes beyond that associated with building pressurization control for most other purposes. It is also likely to increase energy use more than an ASD system. In cases where the building owner is not the party legally responsible for mitigation, the owner may be reluctant to employ the HVAC system for vapor intrusion mitigation control. For these reasons, building pressurization has not been widely used as a vapor intrusion mitigation strategy to date (Folkes 2008).

- *Other Techniques.* Positive sub-slab pressurization, indoor air treatment, and "intrinsically safe building design" may be viable alternatives or supplemental mitigation techniques in certain circumstances.

## Vapor Intrusion Mitigation to Accelerate Redevelopment of a Brownfield Site

**Figure 3.4-A** Gas Vapor Barrier Being Sprayed onto Geotextile Base Fabric (Additional photographs from this project are shown in Figures 3.4-D, 3.4-E, and 3.4-G.)

*Photograph copyright CETCO.*

A big-box retailer on the East Coast wanted to build a store on a brownfield site that had been occupied by a chemical manufacturing plant for more than 50 years. Both the soil and the groundwater were contaminated with a number of VOCs. The primary chemicals of concern for vapor intrusion were xylene, ethylbenzene, and chlorobenzene.

Site remediation included excavation and removal of source-zone soils and air sparging with soil vapor extraction for long-term in situ treatment of the shallow groundwater in the overburden and weathered upper bedrock. Institutional controls were established limiting the future use of the site and requiring site cover.

Air sparging with soil vapor extraction was expected to take several years to reduce concentrations to levels that would allow redevelopment without additional engineering controls for vapor intrusion. Further, the air sparging process promotes volatilization of the chemicals of concern and tends to pressurize the subsurface in the treatment area, so it can temporarily increase the risk of vapor intrusion. Several approaches were considered to mitigate this risk and enable earlier redevelopment. The approach ultimately selected combined a passive sub-slab venting system and spray-applied gas vapor barrier (Figures 3.4-A and 3.4-B) with ongoing monitoring and contingency plans that could be implemented if monitoring results exceeded regulatory action levels.

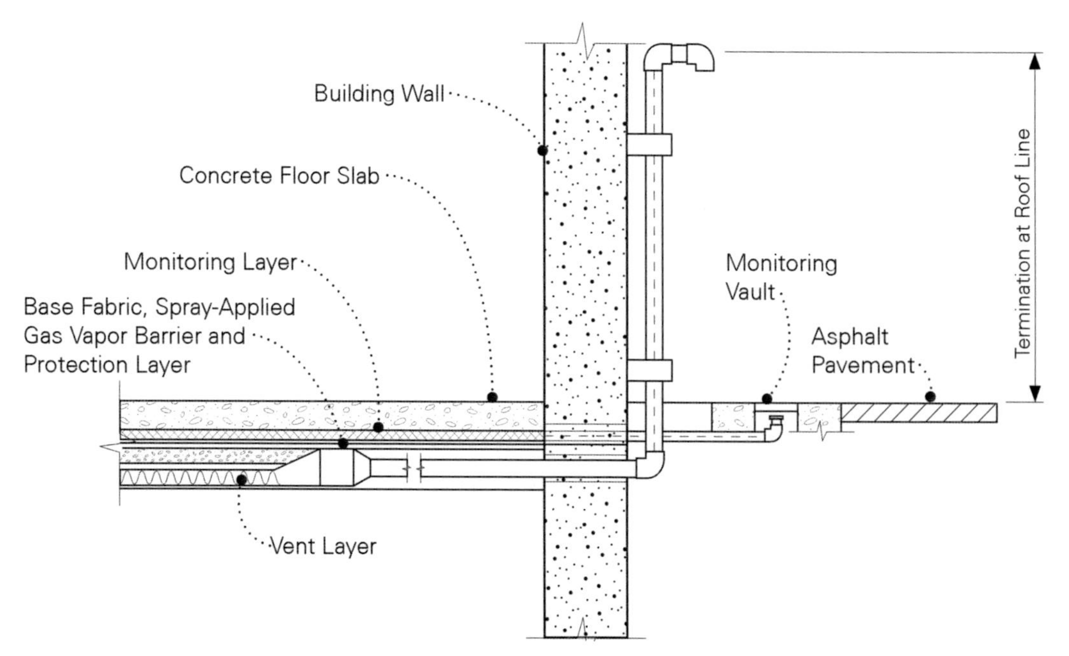


**Figure 3.4-B** Cross Section of Venting Layer, Gas Vapor Barrier, and Monitoring Layer

*Adapted from an image copyright CH2M HILL and The Dow Chemical Company (Rohm and Haas is a wholly owned subsidiary of The Dow Chemical Company).*

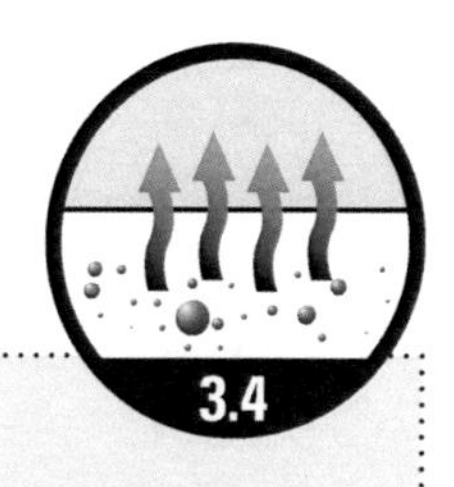


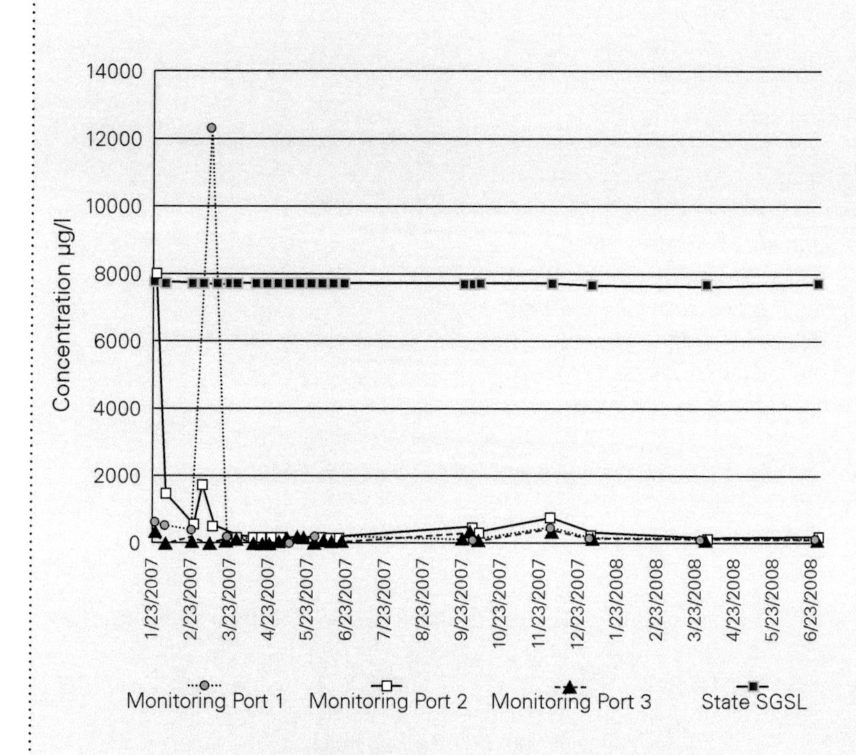


**Figure 3.4-C** Concentrations of Xylene in the Air Samples from the Monitoring Layer (The graph covers an 18-month period starting with the initial samples [Lowe et al. 2009].)

*Adapted from an image copyright CH2M HILL and The Dow Chemical Company (Rohm and Haas is a wholly owned subsidiary of The Dow Chemical Company).*

*Case study courtesy of CH2M HILL (Lowe et al. 2009) and CETCO Liquid Boot Co.*

For this project, monitoring of indoor air was deemed problematic since both the building finishes and some of the products sold by the retailer might emit VOCs, including some of the chemicals of concern. The regulatory agency agreed to allow the remediator to monitor concentrations in a permeable monitoring layer installed above the gas vapor barrier and beneath the concrete foundation (Figure 3.4-B). Concentrations exceeding the state's soil gas screening levels (SGSLs) would trigger confirmatory sampling followed by execution of contingency plans. These included conversion of the passive venting system to an active system, temporary discontinuation of air sparging (while continuing soil vapor extraction), and more frequent sampling. Further contingency plans were available if the more frequent samples remained above screening levels.

Initial vapor concentrations in the monitoring layer were above the state's SGSLs for some compounds (Figure 3.4-C). In response, the monitoring layer was purged to remove any construction-related residuals, the air sparging system was temporarily turned off, soil vapor extraction was enhanced, and the venting layer was converted from passive to active operation by connecting it to the soil vapor extraction system. The frequency of monitoring was increased, and concentrations were observed to fall rapidly to well below screening levels. Since that time, the sampling interval has been gradually increased as concentrations have remained far below action levels.

The vapor intrusion mitigation system was successful in enabling redevelopment of this brownfield site well ahead of the schedule that would have been permitted with soil and groundwater remediation alone.

# Provide Effective Track-Off Systems at Entrances



Dirt and moisture transported into a building on the footwear of its occupants can be a significant source of indoor pollutants because the dirt carries a variety of contaminants and, combined with the moisture, may foster the indoor growth of biocontaminants. Various contaminants of concern (CoC) have been identified in tracked-in dirt (including pesticides). Tracked in dirt and moisture also increases the need for indoor cleaning and thus indirectly degrades IAQ through the unnecessary release of contaminants associated with cleaning activities (see Strategy 5.3 – Minimize IAQ Impacts Associated with Cleaning and Maintenance). The best way to reduce the IAQ impact of these tracked-in pollutants is to develop effective barrier systems. Note that entranceways are also major pathways for outdoor airborne contaminants (including moisture) to reach building interiors and are thus critical zones for building pressurization strategies (see Strategy 2.3 – Maintain Proper Building Pressurization). As part of total entranceway design, track-off systems need to consider and accommodate pressurization strategies, including the use of revolving doors or two-door vestibules.

Barriers to tracked-in dirt begin with the approaches to the building itself and include appropriate selection of landscaping materials and plants. Since pesticides applied outdoors can be readily carried into the building on footwear, well-considered pest control strategies are also important (see Strategy 3.6 – Design and Build to Exclude Pests). To prevent dirt accumulation on footwear, well-designed and well-laid-out walkways using textured paving materials need to be selected. An effective building maintenance and cleaning strategy also needs to be included in the O&M documentation and training (see Strategy 1.5 – Facilitate Effective Operation and Maintenance for IAQ) to ensure that building entranceways are kept clean and, to the extent possible, dry. Any landscaping materials that drop flowers or berries that can be tracked into the building need to be avoided near walkways.

Installation of effective dirt track-off (or walk-off) systems at all entranceways is an essential component of a building's IAQ strategy. Zones such as loading docks, receiving areas, and garage entrances may not have the traffic density of main entrances but can be dominant contributors to tracked-in contaminants within the building and therefore need to be carefully considered.

Proper design of dirt track-off/entry mat systems needs to include specific combinations of mat materials, textures, and lengths. A three-part system is generally recommended: an initial scraper section installed outside building entrances serves to remove loose dirt and water (or snow), stiff-bristled adsorption mats located immediately within the building (also called *trapper* or *wiper* sections) remove additional dirt and moisture via brushing/scrubbing action, and final finishing (or *duster*) mats complete the process by removing particles left after the scraper and adsorption mats.

Final design of the track-off system needs to consider traffic loads and aesthetics as well as local environmental/climate conditions. Systems installed in snowy climates typically require greater lengths of scraper mat, while in rainy locations longer adsorption portions are needed. Muddy locations mandate the need for extended lengths of all three track-off zones.

In addition to decreasing the amount of outdoor contaminants brought into the building, track-off systems also provide the additional economic benefit of increased life expectancy for flooring materials by reducing dirt abrasion of flooring materials.

3.5



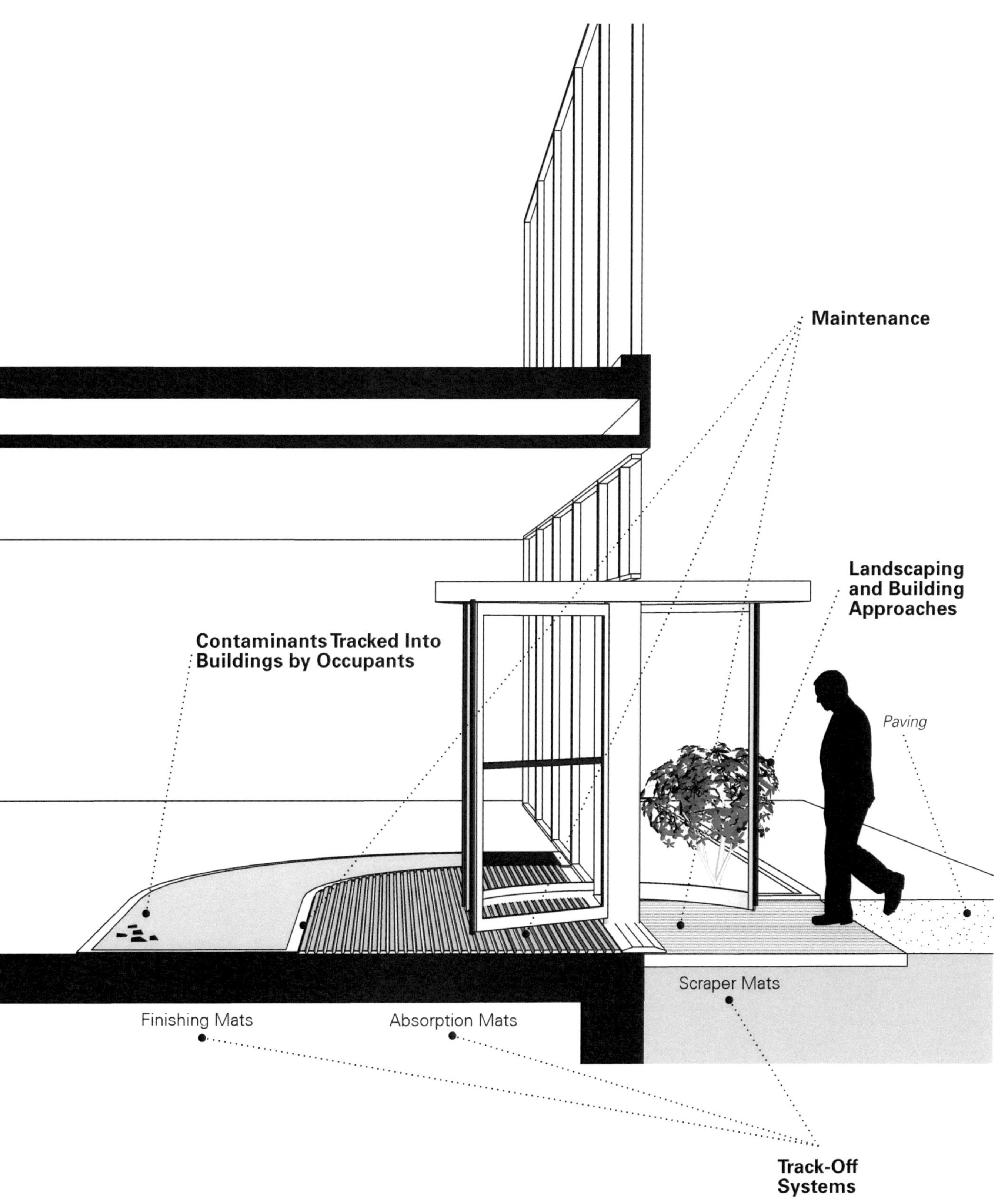
Maintenance
Landscaping and Building Approaches
Contaminants Tracked Into Buildings by Occupants
Paving
Scraper Mats
Finishing Mats
Absorption Mats
Track-Off Systems

## Design Elements Lead to Effective Track-Off System

In the fall of 1996, The H.L.Turner Group Inc. embarked on the design and construction of its new corporate offices in Concord, NH. The intent of the design included many features to enhance IAQ and energy efficiency. Site constraints dictated that the main entrance to the facility (Figure 3.5-A) be located in a predominantly northeastern exposure. Given the normal New England winters, the entrance and flooring system design became an important consideration for fostering good IAQ and ease of maintenance.

**Figure 3.5-A** Entrance Design with Effective Track-Off System Features
*Photograph copyright The HL Turner Group, Inc.*

Key design elements evolved to include:

1. *Heated Walkway.* Thermostatically controlled heat, provided by means of a walkway heat loop supplied by the hydronic system, eliminates the use of sand and ice melting chemicals at the north entrance. Maximum efficiency is achieved by positioning this loop just before the heating water returns to the condensing boilers.
2. *Canopied Entrance.* There is a canopied entrance with the first aluminum track-off grating element located just outside the vestibule entrance.
3. *Vestibule Entrance.* The vestibule entrance has a hard surface surrounding the second track-off grating, this one with a rubber matting cover and a dirt catching pit below. The pit is exhaust-vented to the building exterior.
4. *Removable Track-Off Matting.* The third track-off element is located just inside the vestibule entrance. The length of this matting can be adjusted seasonally.
5. *Entrance Area Flooring.* Hard surface flooring is located in the gallery area between the third track-off mat and the primary flooring.

The three track-off areas and hard surface gallery area provide a total of 30 ft (9 m) of walk-off distance before the primary flooring is reached. The soft non-flow-through textile primary flooring was selected to facilitate dirt removal and easy maintenance. Daily vacuuming includes a central vacuum system that is exhausted to the outdoors after the dust is removed from the airstream.

Results. Carpet dust sampling conducted a few years after occupancy revealed extremely low levels of extractable dust (less than 0.017 oz/ft2 [5 g/m$^2$]) except for the area immediately adjacent to the entryway, where it was 0.020 oz/ft$^2$ (6.1 g/m$^2$). Visible staining of the surface is minimal, requiring only annual extraction to remove surface staining in heavily trafficked corridors.

3.5

# Design and Build to Exclude Pests



Buildings may experience infestations from a variety of creatures. These include an assortment of mammals, insects and arthropods, rodents, birds, and fungi. These creatures can bring about both infectious diseases and allergic reactions in occupants, produce unpleasant noise or odors, cause emotional distress to occupants, damage the building fabric, or bring about the use of pesticides, which results in pesticide exposure to building occupants. Preventing and controlling infestations is therefore of paramount importance.

"Architecture plays a key role [in infestation prevention and control] because design features alone may provide exterior shelter for and/or allow access by pests to interiors" (Franz 1988, p. 260). Building managers report that many new buildings with innovative, energy-efficient designs have pest problems that could have been reduced or avoided with better planning at the design and build stage (PCT 2008; Merchant 2009). In addition, landscape design, construction errors, and poor construction site management can increase the risk of pest colonization of a new building after infestation.

It is possible to design and construct buildings that are resistant to colonization by pests. All colonizing organisms need a point of entry to the building; sources of food and water within or near the building; protected locations where they can eat, rest, and find a mate (called harborage); and passages that allow them to safely move among entry, food, water, and *harborage* areas. Left to their own devices, a population of colonizing organisms will expand until it comes to equilibrium with the available food, water, and harborage. In ecological terms, this is referred to as the *carrying capacity* of the building.

To design a building resistant to colonization requires the following steps:

- *Identify Pests of Concern.* Identify the organisms likely to colonize the building based on its proposed location. For example, American cockroaches, German cockroaches, Norway or roof rats, and house mice are likely in many urban locations.
- *Block, Seal, or Eliminate Pest Entry Points.* In the proposed building design, identify the likely entry routes and seal the building enclosure to prevent pest entry. Examples include gaps around doors and windows, between the foundation and the upper portion of the building, or around utility pipes, conduits, or wires.
- *Reduce Risk of Pest Dispersal Throughout Building.* In the proposed building design, identify the likely passageways pests could use to move freely between food or water resources and harborage and eliminate, block, or seal off these routes. This includes gaps around floor and ceiling joists; penetrations in walls, floors, and ceilings; or openings around shafts and chutes.
- *Reduce Pest Access to Food and Water Resources.* In the proposed building design, identify potential sources of food and water that pests might exploit and take steps to block access to these areas. For example, kitchens and garbage handling areas are likely to provide food for many different organisms.
- *Limit Areas of Potential Pest Harborage.* In the proposed building design, identify and eliminate or block access to areas where pests might find harborage. This includes spaces behind brick veneers or sidings; wall cavities, porches, attics, or crawlspaces; plants or trees planted near the building; and specific architectural features.
- *Provide Access for Maintenance and Pest Control Activities.* Since it may not be feasible to eliminate or seal all potential entry points, passageways, voids, and

# Strategy 3.6

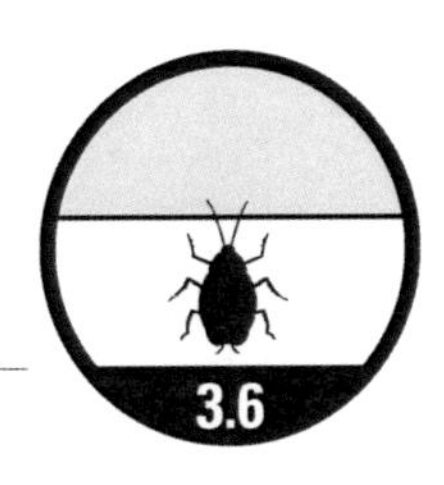


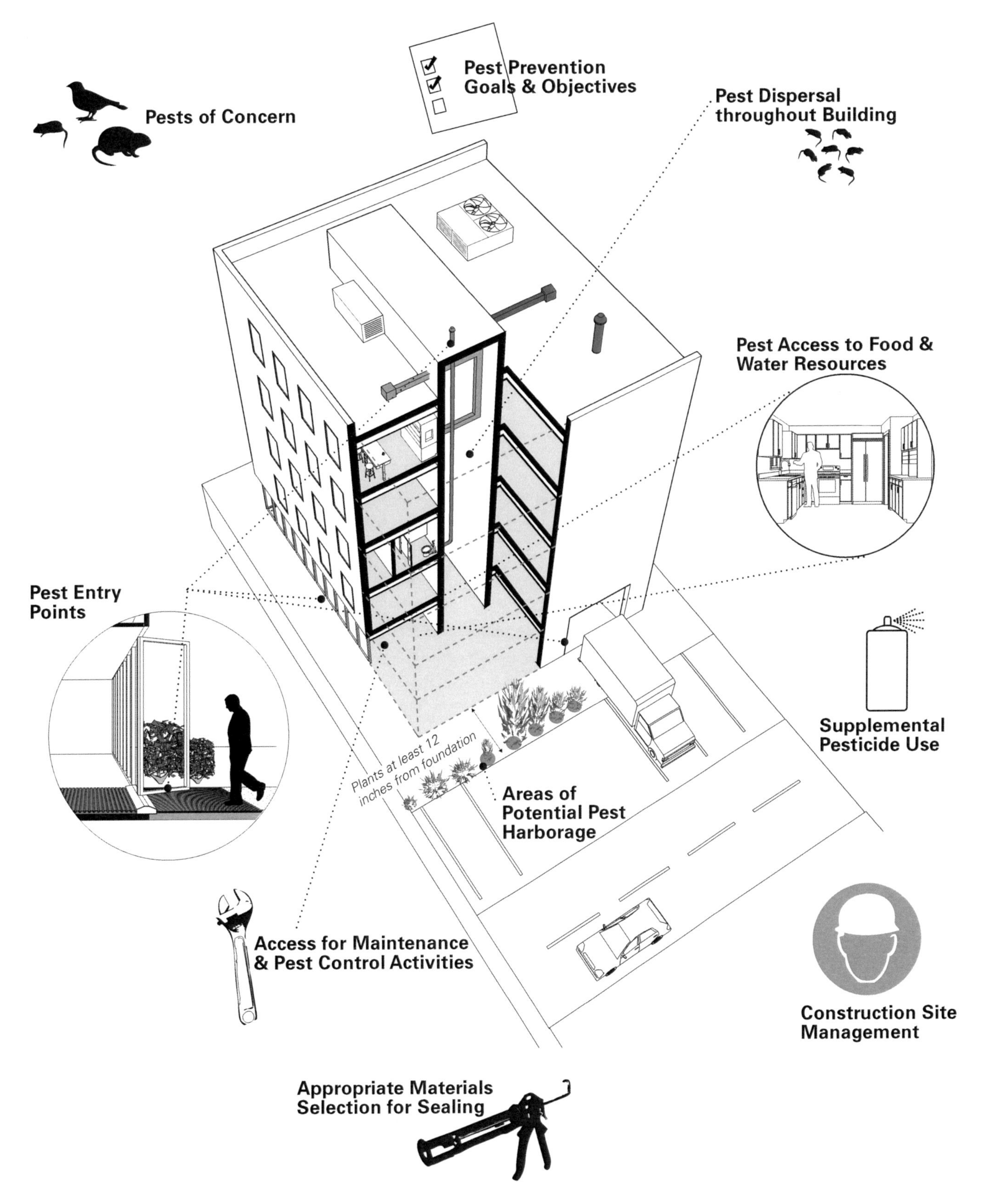


STRATEGY
OBJECTIVE
3.6

harborage sites within a structure, good building design must provide accessibility to such areas for maintenance, cleaning, and possible pest control activities.

- *Appropriate Materials Selection for Sealing.* Sealing or blocking pest access to potential entry points, food and water resources, and harborage sites must be done using materials and methods that are appropriate for the organisms identified as likely colonizers in the neighborhood of the building site.

- *Supplemental Pesticide Use.* It may be necessary to consider the use of pesticides as part of building construction. For example, in some areas of the country, a termiticide or bait system may be needed (or required by law) in order to prevent termite colonization.

- *Construction Site Management.* Since pest infestations can begin as a new building is being erected, pest control needs to be integrated into construction site management to reduce the potential for pest colonization.

## Adding Pest Control Features to Building Design without Adding Cost

Early in the design phase of a housing complex built in New York City from 2004–2006, an IAQ expert recommended incorporating pest control (in addition to other IAQ improvements) into the design of four seven-story low-income multi-family buildings (Figure 3.6-A). Excited about the concept, architect Chris Benedict took on the task of convincing the buildings' owner to approve the plan, which was designed to target rodents, pigeons, and cockroaches.

Pest control features included:

- slab on grade, with all penetrations from below grade for services and plumbing sealed,
- boiler rooms and makeup air louvers placed on the roofs,
- pigeon-resistant lintels over windows,
- boric acid treatment in cavity walls surrounding plumbing chases,
- trash rooms on each floor with the trash chute separated from living space by air barrier construction, and
- floor drains with positive pitch to drains in mechanical rooms where the water could accumulate.

The design team was able to construct the buildings with the pest control features (and many other elements for good IAQ) for the same per-square-foot cost as a typical building without such features. They did this through innovative designs that reduced costs in other ways. For example, for the brick veneer the designers used stainless steel brick ties and created a novel type of window detail to accommodate expansion and contraction instead of installing very expensive steel relieving angles every couple of floors. The window design also eliminates the entry of rainwater without the use of exterior caulking, greatly saving on maintenance costs. The elimination of the relieving angles also means there are no thermal short-circuits in the insulation. Another cost-saving aspect of the buildings is the location of the boiler rooms on the top floor, which greatly reduced the expense of long chimneys and has the added benefits of eliminating potential pest passageways and freeing up area on the lower floors.

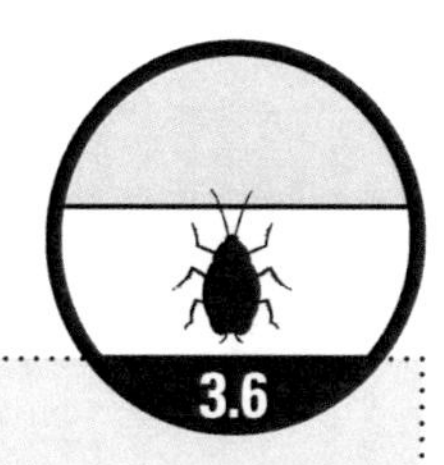


The exterior walls of the building are made with concrete plank and CMU construction, making a durable structure that is relatively easy to seal for pest exclusion (Figure 3.6-B). The CMU makes a vapor barrier, drainage plane, and pest barrier. In addition, individual apartments were compartmentalized to reduce unit-to-unit pest migration. The buildings' manager reports that the pest control measures appear to be very successful, as they have had no pest infestations to date.

**Figure 3.6-A** Exterior of one of the buildings in the complex.

**Figure 3.6-B** Exterior Wall of Parged CMU

*Photographs courtesy of Terry Brennan.*

# Control Moisture and Contaminants Related to Mechanical Systems

Mechanical systems play an important role in providing good IAQ through ventilation, air cleaning, and comfort conditioning. However, since many mechanical systems carry water or become wet in operation, they can also amplify and distribute microbial contaminants. In occupants this can cause building-related symptoms such as nasal and throat irritation and, more rarely, building-related illnesses (BRIs) such as Legionnaires' Disease or humidifier fever. The Strategies in this Objective can help reduce the likelihood of IAQ problems related to mechanical systems.

- Moisture and dirt in air-handling systems provide an environment for microbial growth. Strategy 4.1 – Control Moisture and Dirt in Air-Handling Systems provides techniques to limit rain and snow entry, manage condensate from cooling coils and humidifiers, and keep airstream surfaces clean and dry.
- Condensation on cold piping or ductwork and leaks from piping and fixtures can lead to microbial growth. Strategy 4.2 – Control Moisture Associated with Piping, Plumbing Fixtures, and Ductwork addresses insulation and vapor retarders, including design assumptions and damage protection as well as reduction of piping leaks.
- Periodic inspection, cleaning, and repair of mechanical systems is critical to IAQ but is often hindered by poor access. Strategy 4.3 – Facilitate Access to HVAC Systems for Inspection, Cleaning, and Maintenance addresses equipment location, clearances, and other access issues.
- Legionella can multiply in building water systems such as cooling towers, humidifiers, potable water systems, spas, and fountains. Inhalation of *Legionella* from these sources causes about 18,000 cases of Legionnaires' Disease and 4500 deaths per year in the U.S. Strategy 4.4 – Control *Legionella* in Water Systems addresses the control of *Legionella*.
- One approach that can be used to limit the growth of microorganisms in air-handling systems is ultraviolet germicidal irradiation (UVGI). Strategy 4.5 – Consider Ultraviolet Germicidal Irradiation discusses the state of knowledge regarding UVGI.

Strategies discussed under other Objectives that also help to limit IAQ problems related to mechanical systems include the following:

- Strategy 1.4 – Employ Project Scheduling and Manage Construction Activities to Facilitate Good IAQ
- Strategy 1.5 – Facilitate Effective Operation and Maintenance for IAQ
- Strategy 2.2 – Limit Condensation of Water Vapor within the Building Envelope and on Interior Surfaces
- Strategy 2.3 – Maintain Proper Building Pressurization
- Strategy 2.5 – Select Suitable Materials, Equipment, and Assemblies for Unavoidably Wet Areas
- Strategy 3.2 – Locate Outdoor Air Intakes to Minimize Introduction of Contaminants
- Strategy 7.5 – Provide Particle Filtration and Gas-Phase Air Cleaning Consistent with Project IAQ Objectives

# Objective 4

**4.1** Control Moisture and Dirt in Air-Handling Systems

**4.2** Control Moisture Associated with Piping, Plumbing Fixtures, and Ductwork

**4.3** Facilitate Access to HVAC Systems for Inspection, Cleaning, and Maintenance

**4.4** Control *Legionella* in Water Systems

**4.5** Consider Ultraviolet Germicidal Irradiation

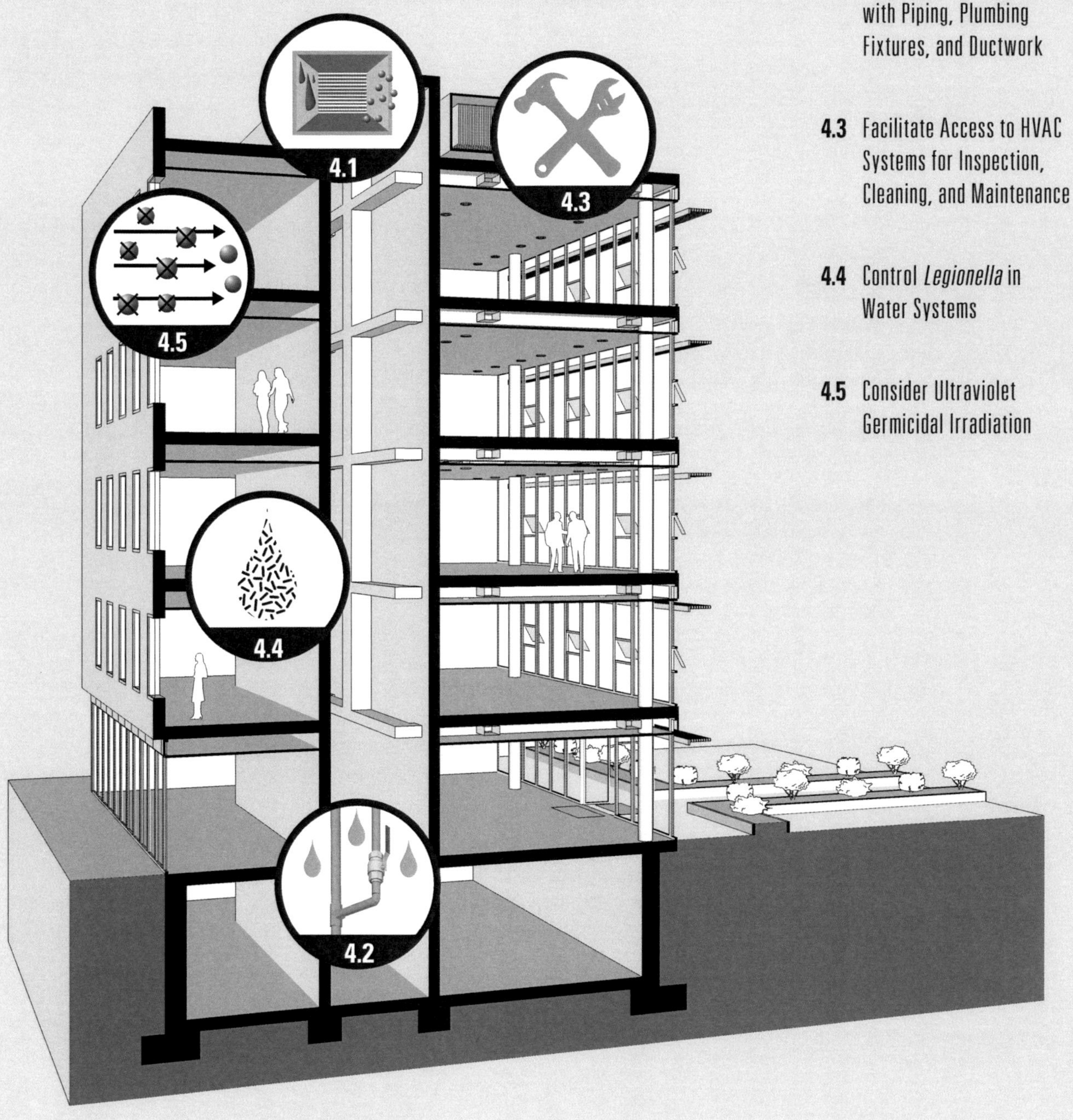

# Control Moisture and Dirt in Air-Handling Systems

Fungi and bacteria are normally present on most interior surfaces in buildings, including on surfaces in HVAC system components. These microorganisms become problematic to IAQ when they amplify or grow on surfaces, sometimes to the point where the growth is visibly obvious. The growth of microorganisms in HVAC systems can result in malodors, building-related symptoms in occupants (e.g., nasal and throat irritation), and in rare cases, building-related illnesses such as humidifier fever and hypersensitivity pneumonitis. Implementation of design strategies that limit moisture and dirt accumulation in HVAC components lessens the risk of microbial growth on HVAC component surfaces.



- *Outdoor Air Intakes and Outdoor Air Inlet Areaways.* Protection against rain and snow intrusion is important. In addition, below-grade outdoor intakes can become accumulation sites for dirt and debris and landscaping pesticides and fertilizers, plus leaves, which are also growth sites for fungi.

- *Filters and Microbial Growth in HVAC Equipment.* Highly efficient filters provide an important tool for reducing the amount of dirt and dust on airstream surfaces that are nutrients for microbial growth under damp-wet conditions.

- *Water Accumulation in HVAC Drain Pans.* Adequate drainage design is critical to limiting microbial contamination. The drain hole for the pan needs to be flush with the bottom of the pan. When the air-handling unit (AHU) is mounted in a mechanical room, it is important to make certain that allowance is made for mounting the drain line at the very bottom of the pan.

- *Moisture Carryover from Cooling Coils.* If the air velocity is too high over part of the coil section (e.g., due to localized accumulation of dirt or poor design), water droplets can and will wet downstream surfaces.

- *Smooth and Cleanable Surfaces.* While microorganisms can grow on smooth but dirty surfaces in HVAC equipment, growth will usually be greatest on porous or irregular airstream surfaces where dust and dirt (nutrient) accumulation is highest. In addition, removal of microbial growth, dirt, and dust from porous or fibrous airstream surfaces can be more difficult.

- *Duct Liners.* It is difficult to achieve a completely clean and dry duct liner that has a fibrous or rough surface over the life of the building with typical, or even above average, airstream filtration. Duct liners with fibrous or rough surfaces present the potential for mold growth since the dirt that accumulates on the surface promotes the retention of moisture and the organic material in the accumulated dirt provides nutrients for mold growth. In addition, it is difficult to remove mold structures, such as hyphae that have grown into fibrous materials.

- *Impact of Humidifier Moisture on Airstream Surfaces.* Water droplets aerosolized from sumps containing recirculated water are heavily colonized by various microorganisms, including actinomycetes, gram-negative bacteria such as *flavobacterium*, and yeasts. It is desirable to use humidifiers that work on the principle of aerosolization of water molecules (absence of carryover of microbes) instead of water droplets (where microbial components may be carried over). Boiler water is not an appropriate source if it contains corrosion inhibitors.

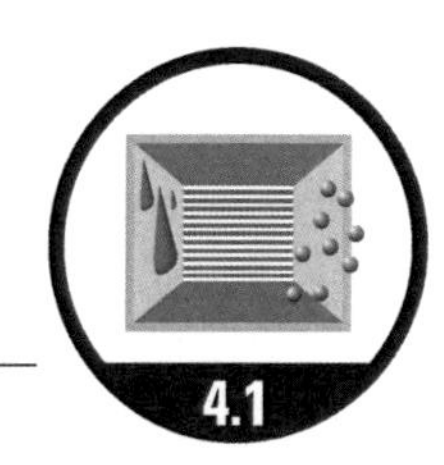
4.1

Moisture Carryover
from Cooling Coils

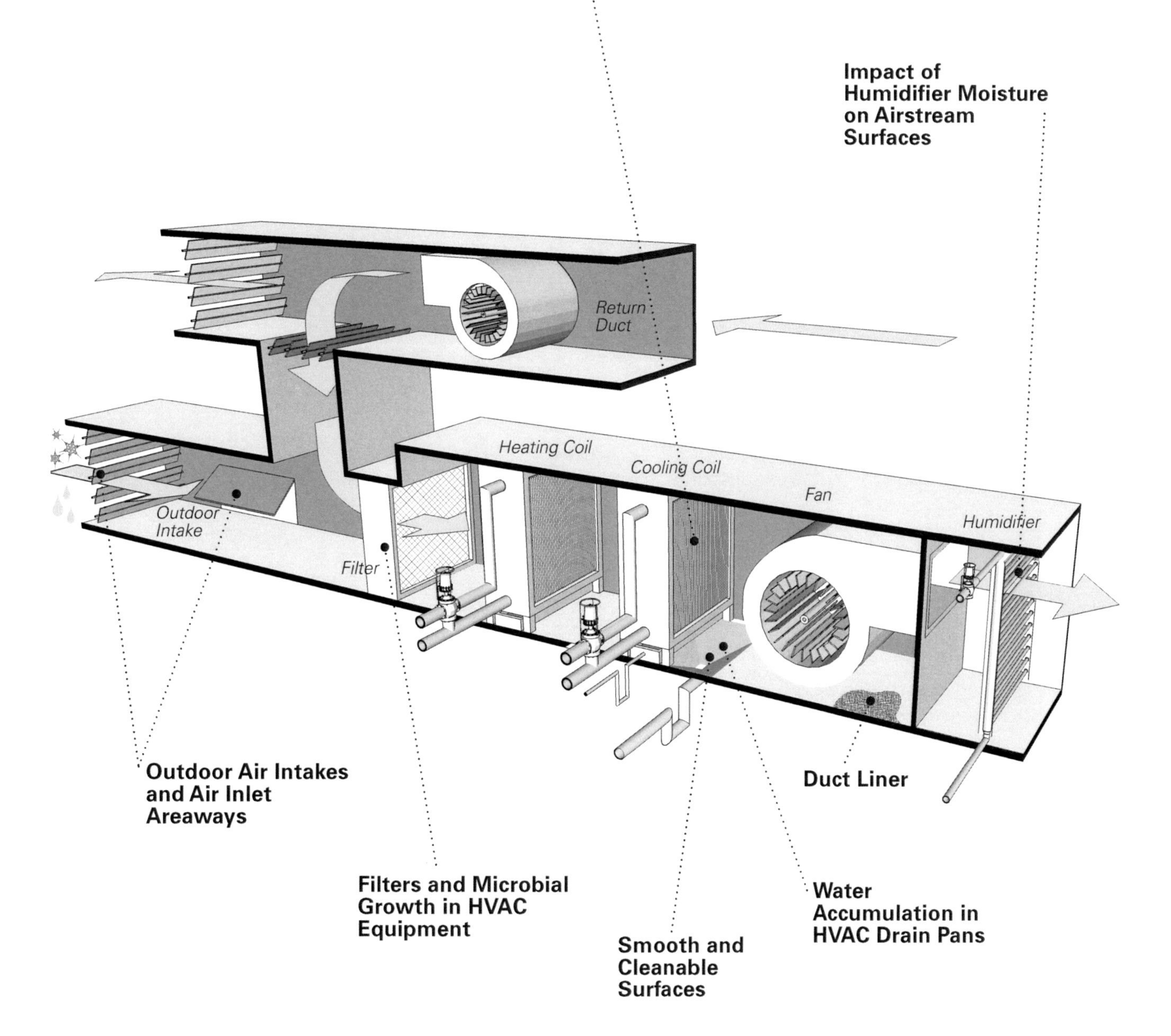
Impact of
Humidifier Moisture
on Airstream
Surfaces
Return
Duct
Heating Coil
Cooling Coil
Fan
Humidifier
Outdoor
Intake
Filter
Outdoor Air Intakes
and Air Inlet
Areaways
Duct Liner
Filters and Microbial
Growth in HVAC
Equipment
Water
Accumulation in
HVAC Drain Pans
Smooth and
Cleanable
Surfaces

## Poorly Designed and Maintained Drain Pan

**Figure 4.1-A** Drain Pan that was Poorly Designed and Maintained
*Photograph courtesy of Phil Morey.*

The AHU shown in Figure 4.1-A was poorly designed and maintained. Access to the pan was achieved only after removal of more than ten fasteners, and the drain pan outlet was not flush with the bottom of the pan. The tan-yellow mass in the pan is a biofilm consisting of a gelatinous mass of fungi, bacteria, and protozoa. This plenum was opened for inspection because of concerns about building-related symptoms and building-related illnesses in the occupied space served by the AHU.

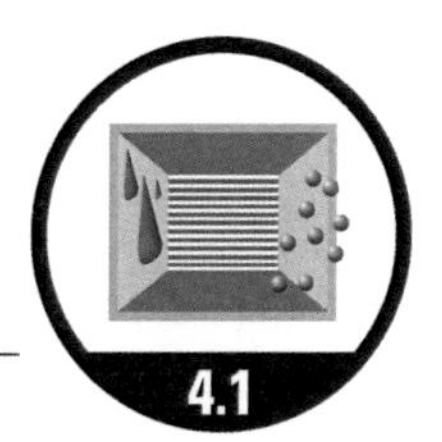
4.1

# Control Moisture Associated with Piping, Plumbing Fixtures, and Ductwork

Mold growth can occur on cold water pipes or cold air supply ducts with inadequate thermal insulation or a failed vapor retarder and result in material damage or significant IAQ problems leading to potential adverse health impacts on occupants. Liquid water from condensation can damage materials nearby such as ceiling tiles, wood materials, and paper-faced wallboard located below or adjacent to the piping or ducts. Leaks from poorly designed plumbing within walls or risers may go unnoticed until damage, including mold growth, becomes evident in occupied spaces. Implementation of design strategies that limit condensation on cold water piping and ducts and that reduce the likelihood of piping leaks hidden in building infrastructure will lessen the likelihood of these potential problems.



## Condensation Associated with High Dew-Point Temperature around Chilled-Water Pipes

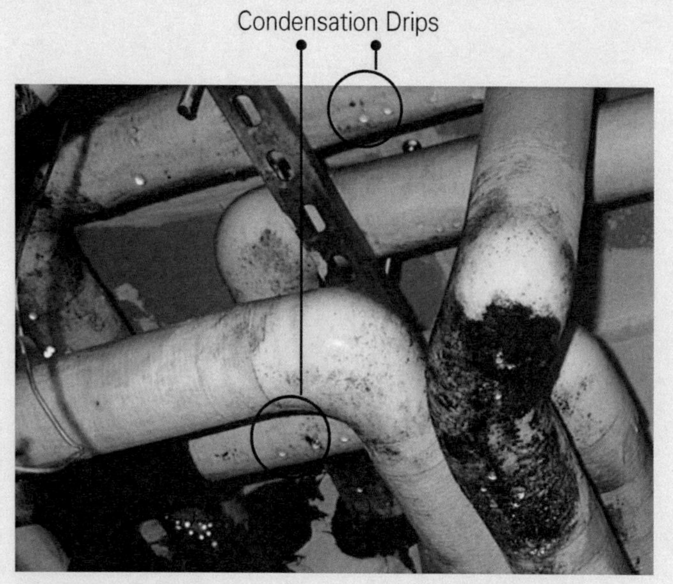


**Figure 4.2-A** Condensation Drips and Mold on Pipe Insulation
*Photograph courtesy of Phil Morey.*

Condensation drips are visible on pipe insulation surfaces in an above-ceiling location, as shown in Figure 4.2-A. Mold growth is present on pipe jacketing. Condensation was associated with the infiltration of warm humid air into the above-ceiling spaces and unexpectedly high dew points in this unconditioned space.

# Strategy 4.2

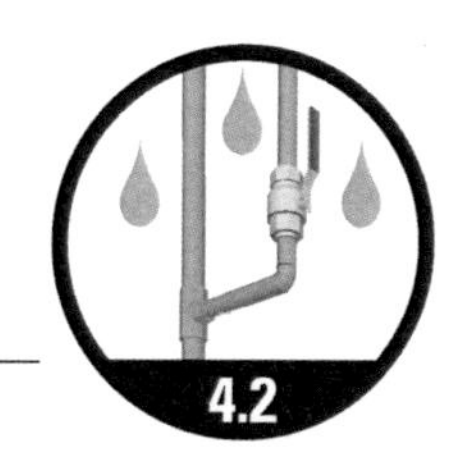
4.2

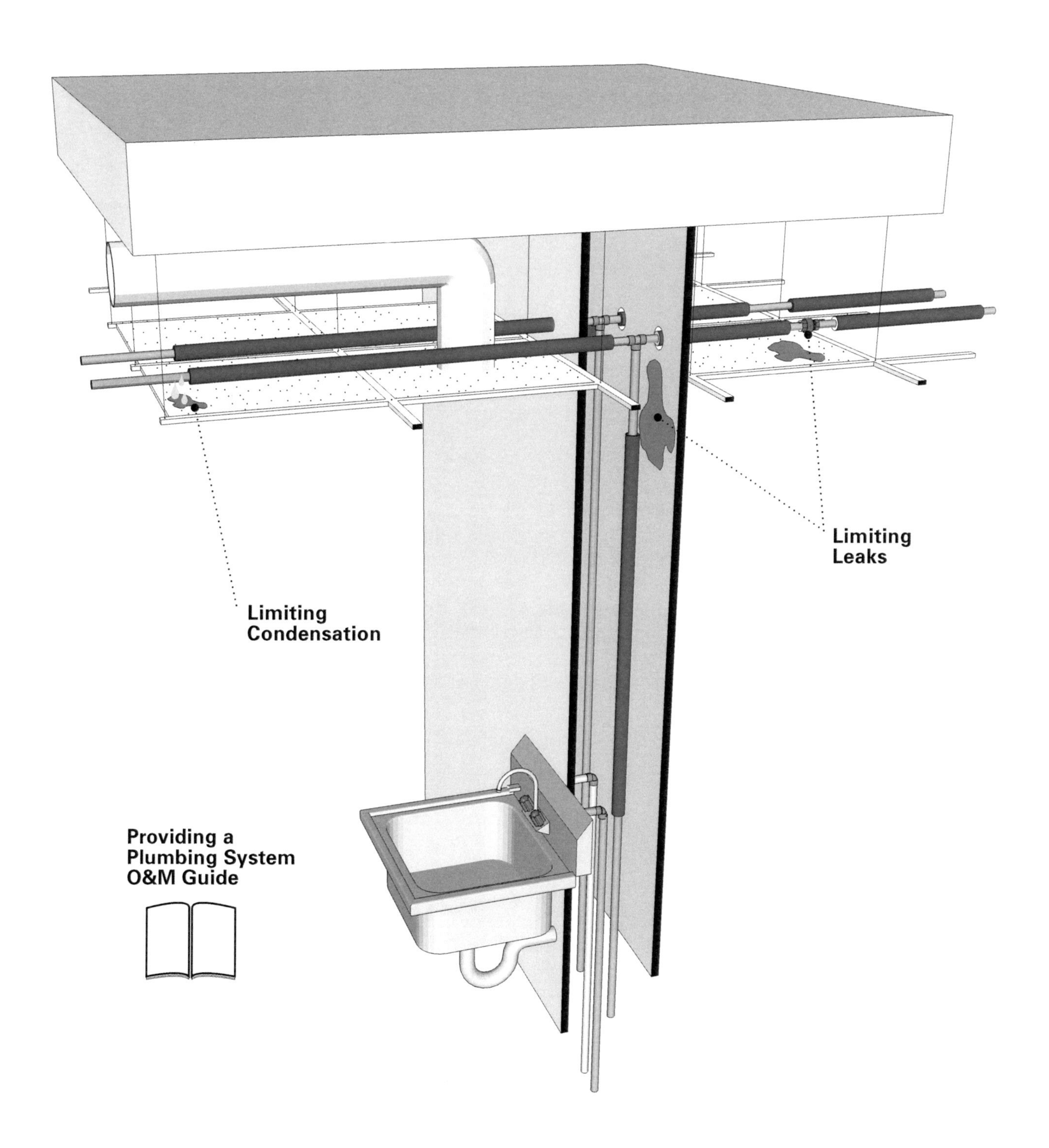
Limiting Condensation
Limiting Leaks
Providing a Plumbing System O&M Guide

# Facilitate Access to HVAC Systems for Inspection, Cleaning, and Maintenance

**Introduction**
**Access in Design Documents**
- Locations that Facilitate Access
- Minimum Clearance Distances
- Critical AHU Components
- Air Distribution System
- System Balancing and Monitoring Access
- Terminal Equipment
- Electrical Code Access Criteria
- Access Door/Panel/View Port Requirements

**Access During Construction**
- Coordination with Trades
- Review of Submittals
- Field Changes
- Monitoring Installations

**Unanticipated Access Requirements**
- Compliance with SMACNA HVAC Duct Construction Standards
- Repeated Access

**References**

Good maintenance of building systems, especially the HVAC system, is a foundation for good IAQ during occupancy. Therefore, during design and construction it is critical that access to the HVAC system for periodic inspection, routine maintenance, and cleaning of the air-handling systems is provided. As HVAC systems become more complex, the ability to access the system, monitor air cleaning, and validate monitor/sensor performance is an increasingly important aspect of building O&M.

### Access in Design Documents

Initial design decisions regarding the type and location of the HVAC system can both limit access and increase maintenance problems. Centralized systems require fewer access points than a network of smaller units. Units installed in inaccessible spaces can provide significant barriers and encourage deferred or neglected inspection and maintenance. Wherever located, the space/room needs to be sized to not only accommodate the equipment but also provide adequate clearance distances that take into account door swings, space for personnel, and movement of tools and materials. Lastly, the design needs to allow adequate space for replacement of major equipment required over the life of the building.

### Coordination with Trades

During construction, the installation of HVAC systems and components needs to be carefully monitored and managed to ensure that clear access is maintained. This includes coordinating with the installation of other building systems, reviewing subcontractor submittals, and assessing the impact of field changes.

### Unanticipated Access Requirements

New situations arise over the life of a building that may require new access points. These points can be either one-time openings that can then be sealed or newly engineered openings that will allow for repeated access. The specific need should be evaluated before any new access points are established.

# Strategy 4.3

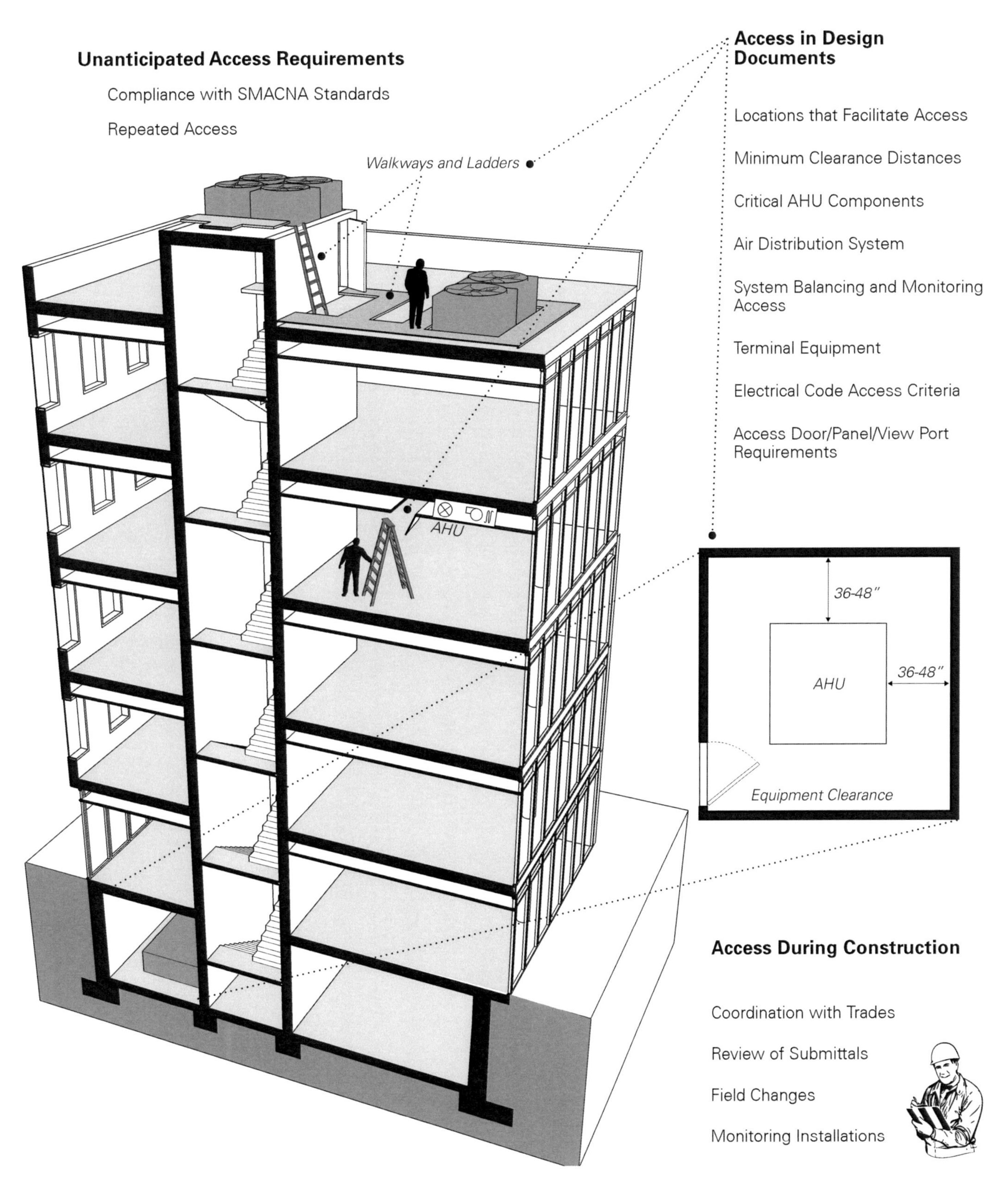


STRATEGY OBJECTIVE 4.3

## Restricted Above-Ceiling Access Compromises Maintenance

**Figure 4.3-A** Access Door to Systems

**Figure 4.3-B** Above-Ceiling System

**Figure 4.3-C** System Location Limits Access

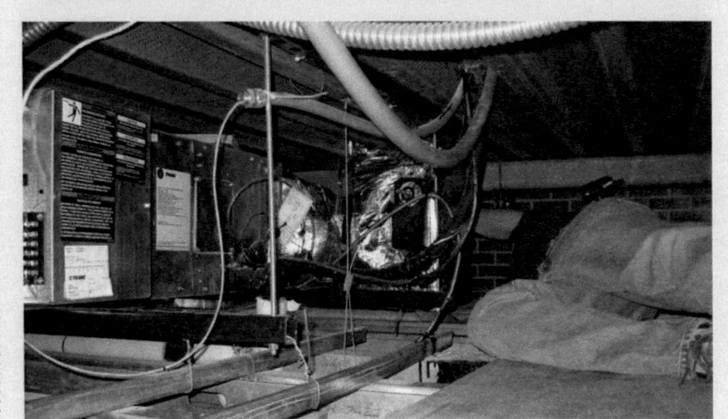

**Figure 4.3-D** All System Components not Readily Accessible

The photographs in this case study (Figures 4.3-A through 4.3-D) illustrate how an above-ceiling installation can significantly compromise access, impede maintenance, and compromise IAQ. In this installation, there were complaints of poor room temperature control and concerns about the rate of outdoor airflow being provided to the space by the fan-coil unit. First, a ladder is required simply to get into the space above the ceiling where the fan-coil unit is located. Notice in Figure 4.3-B that there is no plywood across the metal ceiling joists to support the weight of a technician. The technician had to procure a piece of plywood that would fit through the access door and lay it across the joists in order to access the outdoor air duct for flow measurement and access the unit to verify proper system and component operation. In addition, the narrow plenum space provides a very cramped work area that prevents reasonable access to key components of the system. It was found that the outdoor air damper was not actuating correctly. This allowed too much outdoor airflow to the unit, which did not have the capacity to handle the extreme summer heat and humidity or the extreme winter cold; this caused the space temperature control problems. The balancing dampers for the two outlets (visible in the ceiling near the glass block wall and door in Figure 4.3-A) could not be accessed, which prevented the proper proportioning of the airflow. Maintenance on such installations is frequently ignored or postponed, which can lead to significant operation and IAQ problems.

*Photographs courtesy of Jim Hall.*

# Control *Legionella* in Water Systems

*Legionella* are bacteria normally present in aquatic environments such as rivers and lakes. These bacteria can also be present in man-made water systems, where they can multiply or grow and potentially cause illnesses known as Legionnaires' Disease and Pontiac Fever. Legionnaires' Disease is a lung infection caused by inhalation of mist or water droplets containing the bacteria. Legionnaires' Disease can be fatal. Approximately 18,000 cases of Legionnaires' Disease occur annually in the United States (Squier et al. 2005). Case fatality rates are approximately 20% for community-acquired Legionnaires' Disease and 20%–40% for hospital-acquired disease, which translates into an annual fatality total of 4,000 to 5,000 (Squier et al. 2005; Benin et al. 2002). *Legionella* can grow in most man-made aquatic environments such as cooling towers, potable water systems including showers and sinks, whirlpool spas, humidifiers, vegetable misters, and decorative water fountains.



## Drift from Cooling Tower Travels ~330 ft (~100 m)

**Figure 4.4-A** Cooling Tower Found to be the Source of Legionnaires' Disease in Nearby Building

*Photograph copyright Janet Stout, Special Pathogens Laboratory.*

The drift from the cooling tower shown in Figure 4.4-A travelled approximately 330 ft (100 m) and was implicated as the source of Legionnaires' Disease in three people at a nearby building. The water in the cooling tower basin, which had been treated with biocide and appeared quite clean, was heavily colonized by *Legionella pneumophila* serogroup 1.

# Strategy 4.4

4.4

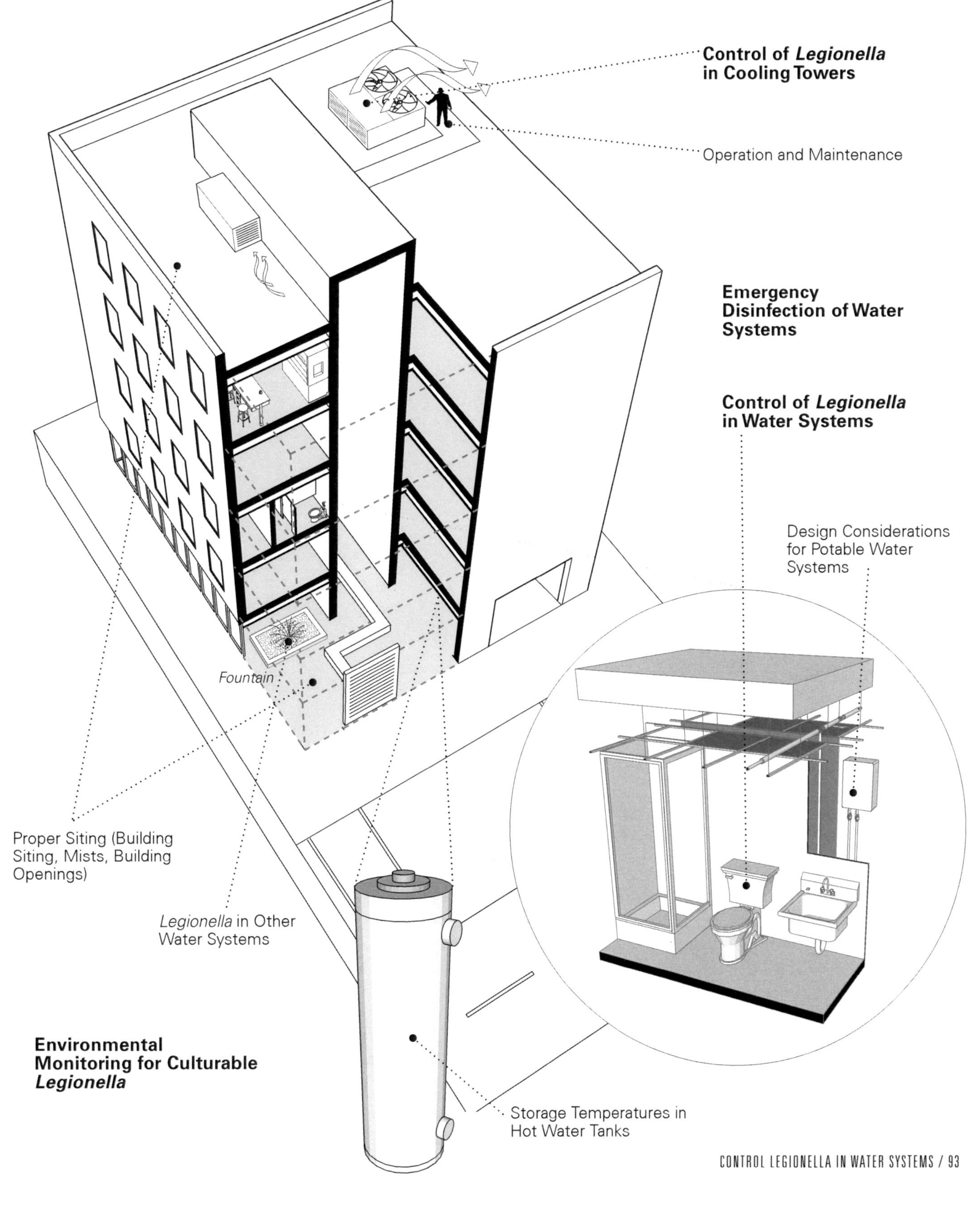


STRATEGY OBJECTIVE 4.4

# Consider Ultraviolet Germicidal Irradiation



Ultraviolet germicidal irradiation (UVGI) has been used successfully for many years to control airborne infective microorganisms such as *Mycobacterium tuberculosis*, the bacterium that causes tuberculosis. Ultraviolet light at all wavelengths but especially at around 265 nanometers (NM) (modern ultraviolet lamps have an optimal discharge at 254 NM) damage the DNA of irradiated microorganisms (Martin et al. 2008). Ultraviolet light emitted and localized in the upper portion of room air (referred to as *upper-air UVGI*) has been used for controlling tuberculosis, especially in poorly ventilated or crowded indoor spaces where other interventions such as increasing HVAC outdoor air ventilation rates or raising filtration efficiency are not practical (Nardell 2002; Nardell et al. 2008). Droplet nuclei[1] from infected occupants in a room can migrate on air currents into the upper (room) air to be inactivated by UVGI from lamps along upper portions of walls. Inactivation of airborne microorganisms depends on both the intensity of the UVGI and the length of time that the particle containing the microbe is irradiated.

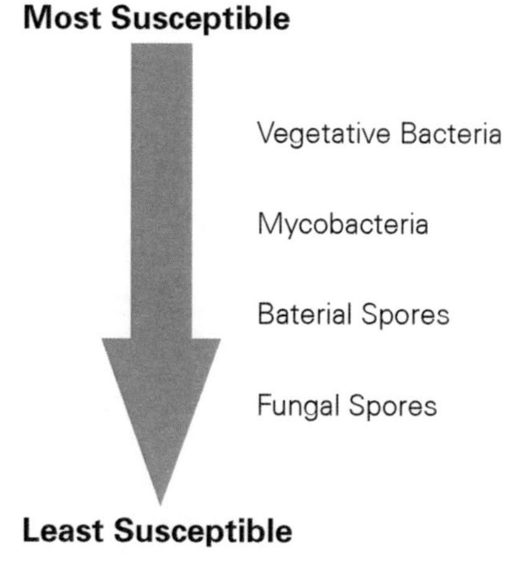


**Figure 4.5-A** Susceptibility of Microorganisms to UGVI
*Adapted from ASHRAE (2008c), Chapter 11, Figure 2.*

In addition to inactivating airborne microorganisms, UVGI directed at environmental surfaces can damage culturable microorganisms present or growing on the surface. Lower-intensity UVGI is effective for surface inactivation because irradiation is applied continuously. UVGI from lamps in AHU plenums has been used successfully to inactivate microorganisms present on airstream surfaces such as on cooling coils and drain pans (Menzies et al. 1999, 2003).

Vegetative bacteria including *Mycobacterium tuberculosis* are most susceptible to UVGI. Fungal and bacterial spores are more resistant to UVGI inactivation (see Figure 4.5-A). It needs to also be noted that viruses are among the most susceptible microorganisms to UVGI (Wells 1943; Perkins et al. 1947).

Studies by Menzies et al. (1999, 2003) have shown that occupants in several office buildings in Montreal reported a reduction in building-related symptoms when UVGI lamps in AHUs were turned on (as compared to time periods when the lamps in the same AHUs were deactivated.) Environmental microbiology tests in buildings studied by Menzies et al. determined that there was a significant decline in culturable bacteria and fungi on irradiated surfaces on cooling coils and in drain pans. However, these intervention studies did not find any decline in culturable fungi or endotoxins in workplace (office) air when UVGI lamps in AHUs were turned on. Air sampling in offices when UVGI lamps were turned on did show a slight, but not significant, decline in culturable bacteria on blood agar medium (not optimal for environmental bacteria). Thus, while Menzies et al. did find a significant decline in building-related symptoms associated with use of UVGI in

[1] Small particles originating from the human respiratory tract; as water evaporates, the particles become smaller and remain airborne because of their small aerodynamic size.

# Strategy 4.5

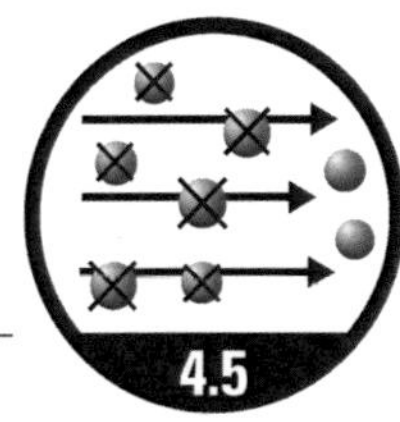

4.5


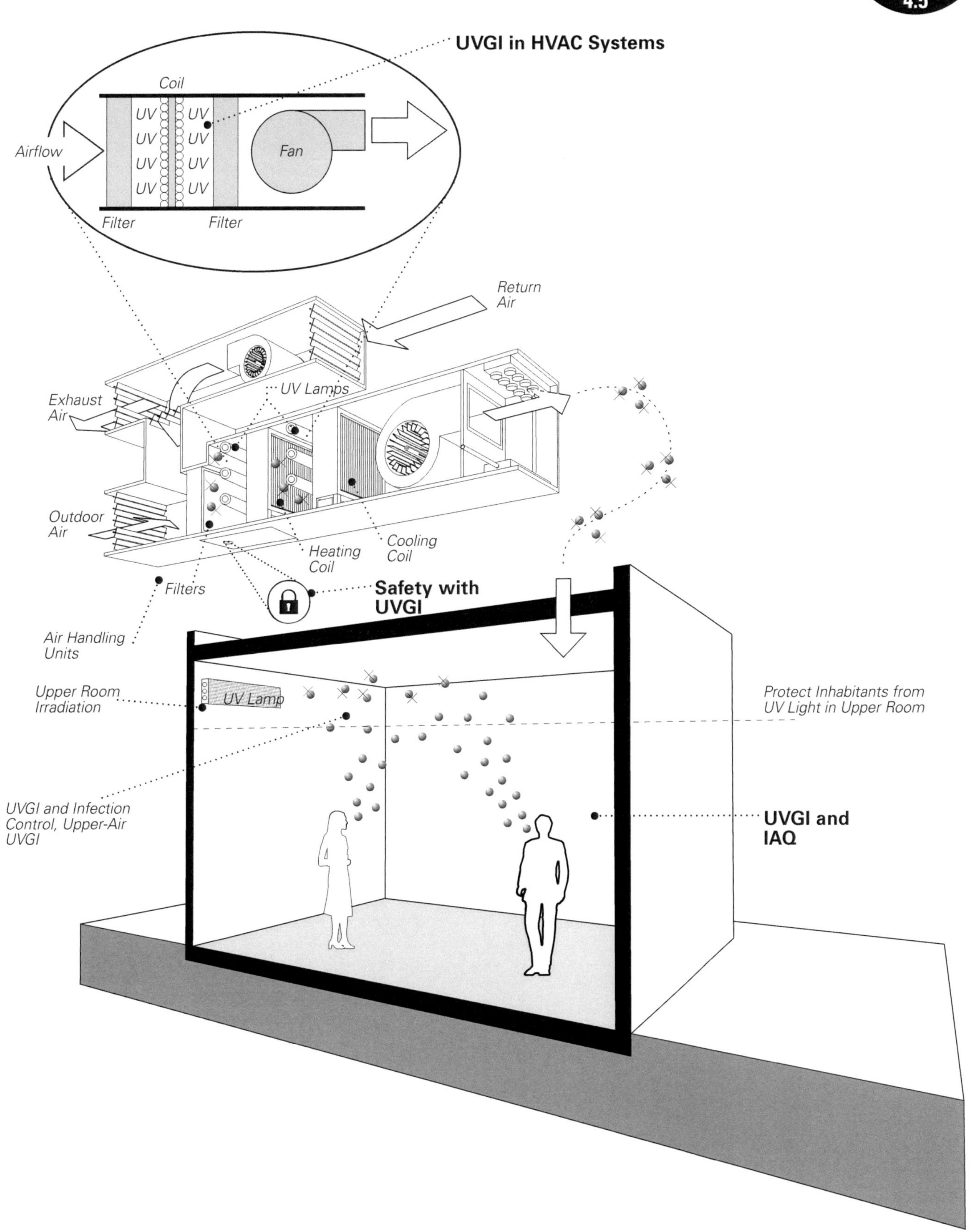

UVGI in HVAC Systems
Coil
UV
Airflow
Fan
Filter
Filter
Return Air
UV Lamps
Exhaust Air
Outdoor Air
Heating Coil
Cooling Coil
Filters
Safety with UVGI
Air Handling Units
Upper Room Irradiation
UV Lamp
Protect Inhabitants from UV Light in Upper Room
UVGI and Infection Control, Upper-Air UVGI
UVGI and IAQ

AHUs, significant declines in airborne levels of culturable fungi and bacteria as well as endotoxins were not detected in the office workplace.

Studies by Bernstein et al. (2006) in Cincinnati homes with asthmatic children showed an association between decline in airway hyper-responsiveness and use of UVGI in home ventilation systems. However, significant declines in concentrations of airborne fungi and bacteria associated with UVGI intervention were not detected. Thus, while both Menzies et al. (1999, 2003) and Bernstein et al. did find positive benefits associated with use of UVGI in HVAC systems, the environmental cause(s) for the reduction in building-related symptoms remains obscure and a topic for future research.

The designer needs to be aware that the use of UVGI lamps in AHUs, ductwork, and upper air requires careful attention to safety considerations to prevent inadvertent exposure of people to ultraviolet light. For example, lockout/tagout procedures are necessary to prevent accidental turning on of UVGI lamps when facility maintenance personnel are working in AHUs. Refer to Chapter 16 of the *ASHRAE Handbook—HVAC Systems and Equipment* (ASHRAE 2008c) for a comprehensive review of safety considerations associated with the use of UVGI in buildings. Well designed upper-air UVGI systems have been used safely for many years (Nardell et al. 2008).

STRATEGY OBJECTIVE 4.5

# Limit Contaminants from Indoor Sources

Many building materials, finishes, and furnishings emit compounds that can cause discomfort, irritation, or other more serious health impacts. These include organic compounds that can cause health effects ranging from eye, nose, and throat irritation to headaches and allergic reactions to organ damage and cancer. The occupant complaints, lost productivity, and absences that can result can lead to additional costs for IAQ investigations, material replacement, and litigation. Some emissions may be relatively benign themselves but react with other compounds in the air, such as ozone, to form secondary products that are more irritating or harmful. Materials, finishes, and furnishings that are difficult to clean may contribute to IAQ problems by necessitating use of strong cleaning agents. While scientific understanding of these issues is still evolving, the Strategies presented here provide practical means to limit their IAQ impacts based on current knowledge.

- Selecting appropriate materials, finishes, and furnishings reduces the likelihood of emissions-related IAQ problems. Strategy 5.1 – Control Indoor Contaminant Sources through Appropriate Material Selection provides background information, describes strengths and weaknesses of emission data sources and rating systems, and contains succinct information and recommendations for a dozen priority product categories.
- Sometimes it is difficult to avoid use of certain materials and products. Strategy 5.2 – Employ Strategies to Limit the Impact of Emissions outlines steps that can be taken to limit the impact of unavoidable emissions, including the use of emission barriers, material conditioning, in-place curing, delayed occupancy, building flush-out, and short-term use of gas-phase air cleaning.
- Cleaning agents and processes can have detrimental effects on indoor air. While this Guide does not address O&M, it does address design to facilitate O&M. Strategy 5.3 – Minimize IAQ Impacts Associated with Cleaning and Maintenance addresses selection of easily cleaned materials and finishes, provision for proper storage and handling of cleaning materials, inclusion of cleaning protocols in O&M documentation and training, and other steps to reduce the IAQ impacts of cleaning.

This Objective focuses on Strategies to reduce indoor contaminant sources. Additional Strategies to deal with contaminants generated indoors include capture and exhaust, filtration and air cleaning, and dilution ventilation. These are discussed under the following:

- Objective 6 – Capture and Exhaust Contaminants from Building Equipment and Activities
- Objective 7 – Reduce Contaminant Concentrations through Ventilation, Filtration, and Air Cleaning

# Objective 5

**5.1** Control Indoor Contaminant Sources through Appropriate Material Selection

**5.2** Employ Strategies to Limit the Impact of Emissions

**5.3** Minimize IAQ Impacts Associated with Cleaning and Maintenance

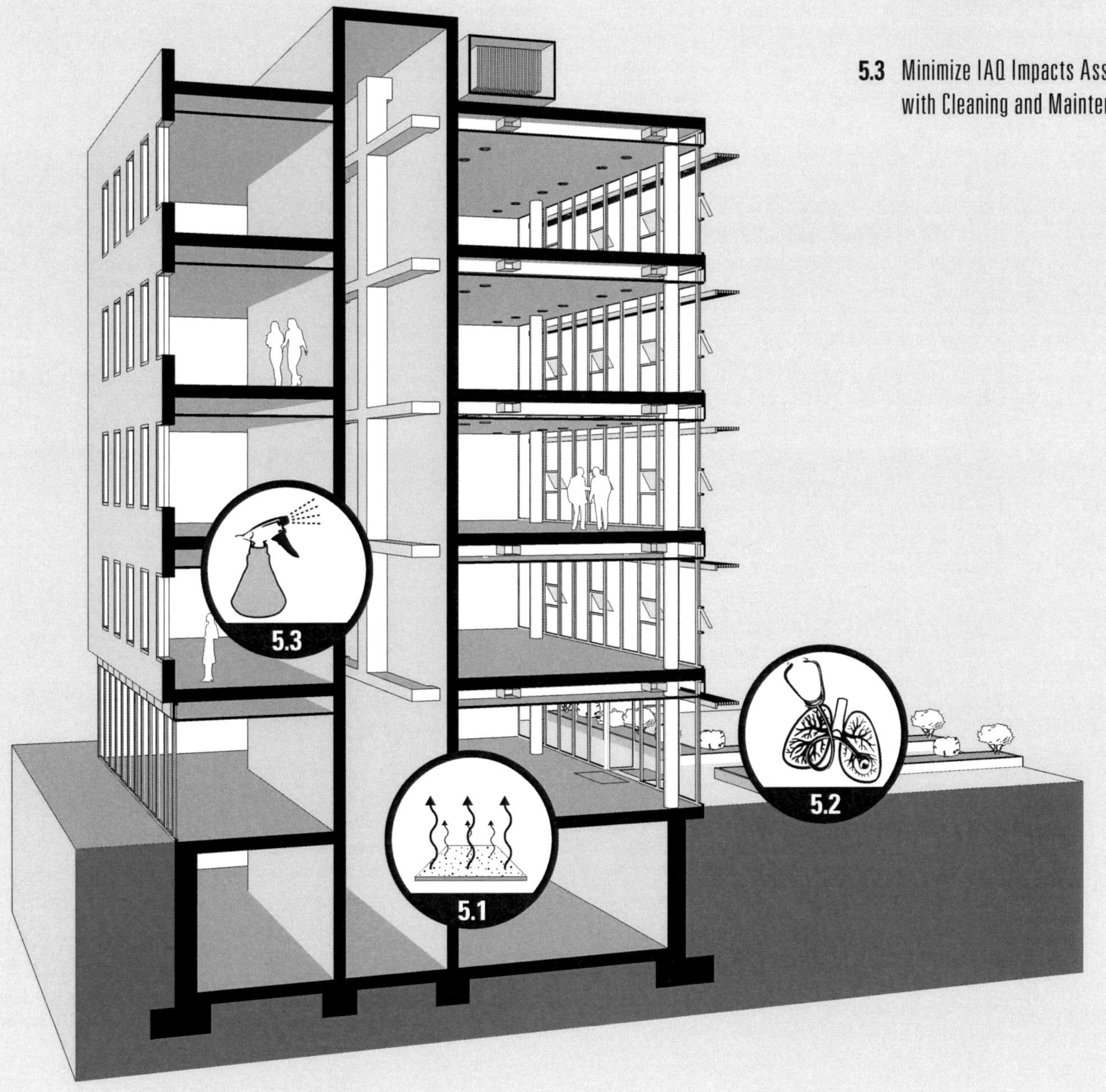


STRATEGY OBJECTIVE 5

# Control Indoor Contaminant Sources through Appropriate Material Selection



Recent advances in the sampling and analysis of indoor contaminants and in toxicology and indoor chemistry have contributed to a greater understanding of the nature and impacts of the pollutants that affect building occupants. In parallel, advancements in the techniques used to determine the emissions (or *off-gassing*) properties of materials and products used in building construction, finishing, and furnishing have enabled us to more clearly see their chemical "fingerprints" on indoor environments and thus their impact on IAQ.

Provision of good IAQ requires coordination of many aspects of building design, and a practical first step is problem avoidance through careful selection of materials with minimal emission of irritating or harmful compounds. This form of source control is an effective means of preventing IAQ problems while reducing the need to dilute avoidable contaminants through costly ventilation. Thus, building designers, in addition to specifying material structural, fire, and moisture (and mold) resistance properties, need to also carefully consider the chemical emission characteristics of materials. Long-term durability, maintenance, and cleaning requirements also have significant impacts on IAQ and need to therefore be included in material specifications. These aspects of material selection need to be considered in the midst of pressure to adopt "green" products that may or may not adequately consider IAQ impact as a component of environmental sustainability.

Rating systems for assessing the chemical emissions of products are still evolving. Product labels describing emission properties provide far more information than that given by content-based product labels, which merely report the percent by weight of VOCs. Within emissions-based systems, the total volatile organic compound (TVOC) emission rate, while still widely reported, is increasingly recognized as a poor indicator of the true impact of any given material. This is largely due to the great range in irritant, odor, and toxicological impact of individual VOCs: some have significant impacts at relatively low levels, while others may be relatively harmless at high concentrations. Long-term emissions of SVOCs such as phthalates, pesticides, and flame retardants are now recognized as important factors in IAQ problems and need to be considered in the evaluation of material properties.

Actual emission rates vary significantly over time: for a given product, emissions of some chemicals decay rapidly (within hours or days), while others may release contaminants at nearly constant rates for many months. The acute or long-term impacts of materials can thus be dramatically different and need to be factored into product assessment. Formation of secondary products through indoor chemistry reactions may have real impact on IAQ; thus, elimination or reduction of the primary reactants (such as terpenoids) could be considered by advanced labeling systems on which designers can base material selection decisions.

In evaluating emissions impacts, materials need to be considered as parts of systems whenever possible. For example, carpeting is not independent of cushions, adhesives, or subfloors. Wallboard requires primer and paint. Emissions from a system may be markedly different than those from its individual constituents.

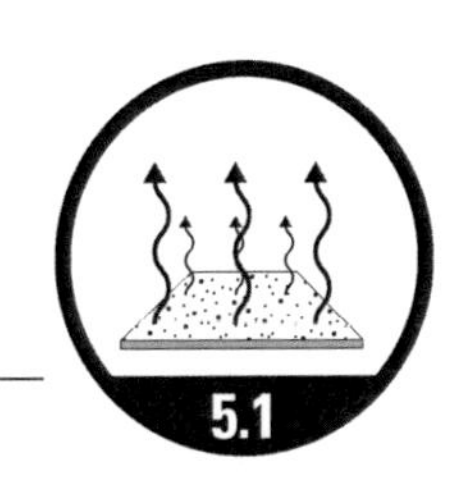


**Contaminant Emissions: Basic Concepts**

- VOCs—Total vs. Target: Irritancy, Odor, and Health Impact
- Semi-Volatile Organic Compounds (SVOCs)
- Indoor Chemistry - Secondary Emissions
- IAQ Guidelines, Standards and Specifications
- Shades of Green - Environmentally Preferred Products
- Product Information - Composition vs. Emissions
- Emissions Behavior

**Emissions Data: Available Information**

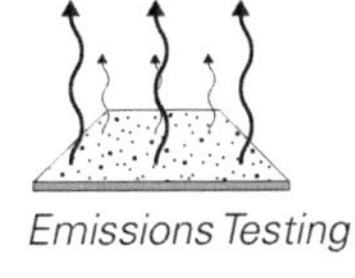

*Emissions Testing*

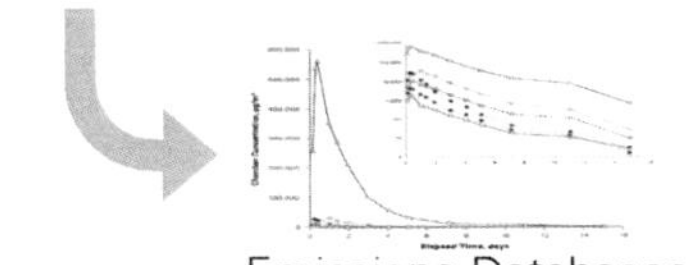

Emissions Databases

Manufacturer Supplied Information: MSDSs

Labels: Emissions-Based

Labels: Content-Based

**Priority Materials/ Finishes/Furnishings**

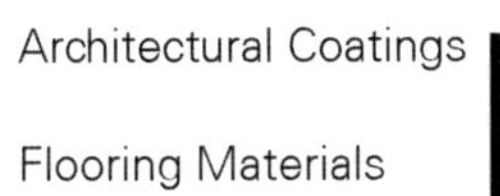

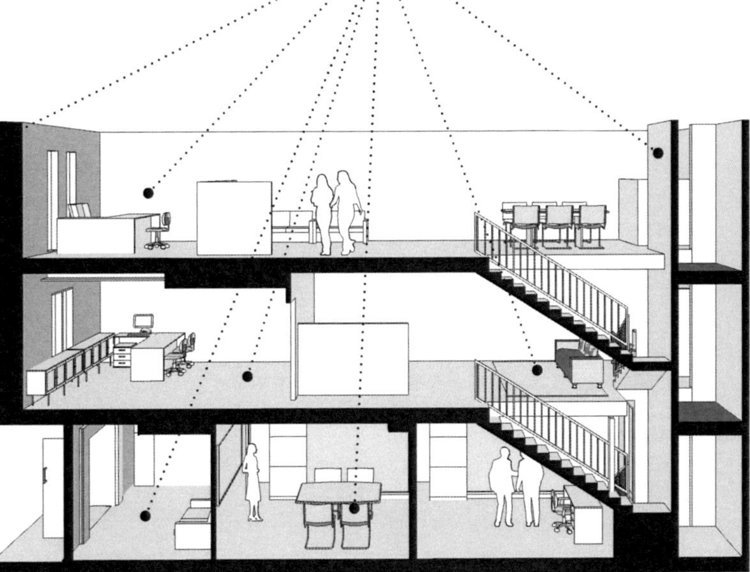

Architectural Coatings

Flooring Materials

Ceiling Tiles

PVC Materials

Insulation Materials

HVAC Components

Caulks, Sealants & Adhesives

Porous of Fleecy Materials

Flame-Retardant Materials

Structural Materials

Office Furniture Systems

Office Equipment

Composite Wood / Agrifiber Materials

Local environmental conditions may influence contaminant release. For example, materials subjected to relatively high temperatures (possibly through solar gains) or high humidity may have increased emissions.

IAQ guidelines and standards for specific contaminants are currently sparse. Occupational regulations for air quality do not directly apply to nonindustrial buildings due to differences in the levels and compositions of the contaminant species as well as the nature of the building occupants. In the absence of legislated limit values, emission labels need to rely on guidance-level information. Building designers need a basic understanding of the key issues related to emissions labeling in order to effectively specify building materials, finishes, and furnishings. In the absence of detailed guidance, several basic recommendations can be made concerning material selection for the diverse range of products that go into building design (Schoen et al. 2008).

In general, the following recommended strategies will assist in selecting low-emitting materials for building design:

- Require submission and review of material composition (VOC contents or, preferably, detailed emissions properties) as condition of acceptance for project (ensure supplier receives detailed information from manufacturer) prior to material selection.
- Where product-specific emissions data are not available, limit usage of products/materials generally known to have higher contaminant emissions, including unfinished composite or engineered wood products; oil-based architectural coatings and paints; and caulks, sealants, and adhesives. Specify use of low-emission resins if required during product manufacture.
- Specify and use products with low-formaldehyde emissions.
- Limit use of porous/fleecy materials including carpeting, fabrics, and upholstery to reduce sink effects and facilitate cleaning.
- Select materials that are durable and low maintenance and have easily cleanable surfaces. Require detailed installation, maintenance, and cleaning instructions as part of the material specification process. Verify that product installation practice conforms to project specification. Ensure that detailed maintenance and cleaning instructions are delivered to building owner/operators (refer to Strategy 5.3 – Minimize IAQ Impacts Associated with Cleaning and Maintenance for additional guidance on cleaning and maintenance that will reduce the IAQ impact of these activities).
- Avoid use of polyvinyl chloride (PVC) based flooring materials in contact with damp concrete that may, through hydrolysis, result in the release of undesirable (secondary) emissions.
- Limit the use of lining materials on interior surfaces of ventilation ducts (see Strategy 4.1 – Control Moisture and Dirt in Air-Handling Systems for guidance on HVAC ducting design to control noise). Use low-emission cleaning agents to remove any residual oils on the interior surfaces of ductwork prior to installation. Immediately following manufacture, seal all duct openings and store in a dry location. Remove seals only just prior to installation to prevent contamination during construction.
- Fully identify each material or product in the project specifications. Prepare a Schedule of Materials, identifying each material by a unique name and symbol that need to appear on project plans.
- Review available emissions information for all substitutions prior to approval; confirm delivered products meet specifications.

Despite best efforts to avoid materials with high contaminant emissions, the use of certain materials and products with moderate to high emissions may still be necessary, depending on building use and function. Additional techniques will therefore be required to limit the effects of material emissions on indoor air. Refer to Strategy 5.2 – Employ Strategies to Limit the Impact of Emissions for further details.

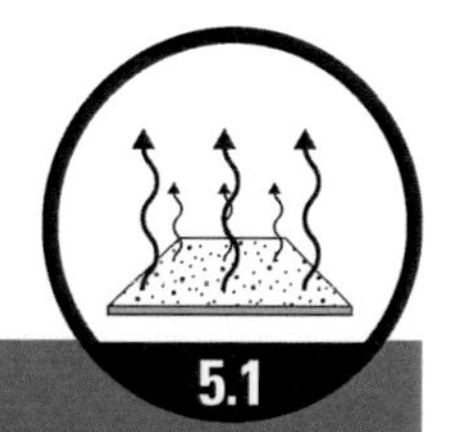


## Selection of Low-Chemical-Emission Materials Leads to Reduced Contaminant Levels in Office Environment

During construction of a new office building, the specification of building materials and furnishings for eight floors to be occupied by one client was made with particular regard for IAQ impact. Construction materials employed on these floors that were carefully screened included insulation, particle boards, wall coverings, paints (latex), stains and varnishes, cabinets, sealing and spackling compounds, glues and adhesives (water-based), tile grout, and plasters and cements. Furnishing specifications included the use of low-formaldehyde fabrics and continuous filament carpeting (to reduce particle shed). Systems furniture and work stations, draperies, and ornamental fabrics were also selected based on IAQ impact considerations.

The contractor employed to construct these eight office floors also simultaneously constructed adjacent floors that did not require similar IAQ specifications for materials and furnishings. These "typical" floors were indistinguishable from the other eight floors in terms of appearance, furnishings, and space usage. Post-occupancy air sampling conducted in the building (Figure 5.1-A) revealed that VOC levels on the low-emission floors were approximately 50%–75% below those found on the conventionally constructed floors.

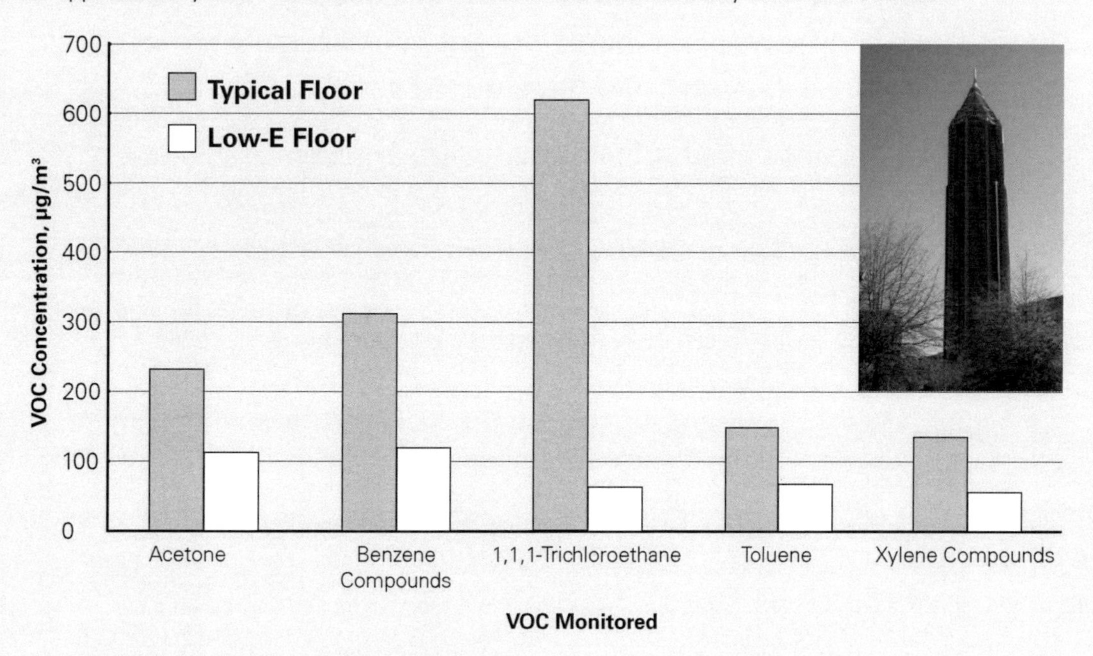


**Figure 5.1-A** Monitored VOC Levels
*Data source: Milam (1994). Inset photograph courtesy of H.E. Burroughs.*

STRATEGY OBJECTIVE 5.1

## Capital Area East End Complex—Sacramento, California

**Figure 5.1-B** Capitol Area East End Complex
*Photograph courtesy of Leon Alevantis.*

The Capitol Area East End Complex (CAEEC) in Sacramento, California, is a five-building sustainable office complex built in 2002–2003 with a total area of 1,500,000 $ft^2$ (140,000 $m^2$) (Figure 5.1-B). Emissions testing of the majority of the interior finishing materials was required per *Section 01350, Special Environmental Requirements Specification* (CIWMB 2000). Details of the project are available at www.eastend.dgs.ca.gov/AboutTheProject/default.htm.

Overall, concentrations of the common chemicals measured at the CAEEC shortly after initial occupancy and for several months thereafter were comparable to those reported in the EPA Building Assessment Survey and Evaluation (BASE) study (EPA 2008d) with only few chemicals at the CAEEC being higher. The BASE study measured contaminant concentrations in buildings that were at least seven years old. In contrast, the CAECC results were collected from a newly constructed building (at the time when emissions from building materials and furnishings are expected to be at their peak). The finding of comparable levels of contaminants in the two studies indicates that careful selection of materials leads to reduced exposures to indoor contaminants, especially during early occupancy of buildings.

The CAECC study shows that requiring emissions testing from manufacturers helped achieve better-than-average IAQ. The concentration targets established for this project were not exceeded in the majority of the locations. Therefore, as expected, careful selection of building materials during a building's design appears to result in lower concentrations of VOCs during the initial months of a newly constructed building (Alevantis et al. 2006).

## EPA Waterside Mall—Washington, DC

This case study shows the costs that can occur when emissions are not considered during material selection.

The first phase of the EPA Waterside Mall, a mixed-use building, was completed in 1970, and the building was occupied by EPA in 1971. Additions were made in the 1980s, including a major renovation in 1987 that included the installation of 243,000 $ft^2$ (22,600 $m^2$) of new carpeting.

Of 3700 employees who responded to a 1989 survey, 880 reported health effects. Although the problem was likely due to a number of factors (including HVAC system inadequacies, occupant crowding, and the density of the office equipment), employees attributed many of the health effects to the new carpeting, which was eventually replaced with a low-odor alternative.

The estimated cost of the replacement was approximately $4 million, including carpet replacement, HVAC renovations, IAQ investigations, sick leave, labor to address IAQ issues, compensation claims, etc., plus litigation costs.

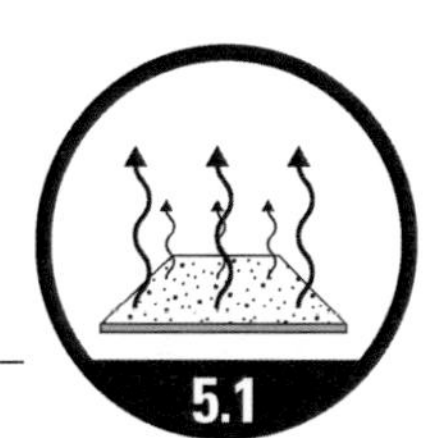
5.1

# Employ Strategies to Limit the Impact of Emissions



The first and most effective means to reduce the impact of material emissions is to employ selection strategies that limit the entry of high-emitting materials into the building (i.e., source control; see Strategy 5.1 – Control Indoor Contaminant Sources through Appropriate Material Selection). The products used in building construction and furnishing will still emit some level of contaminants, and certain materials with relatively high emissions may be unavoidable due to a lack of alternative products or due to emission-generating activities. To reduce the negative impact of these materials/activities on the indoor environment, the building design team needs to consider a variety of alternative strategies.

### Control of Emissions through Use of VOC Barriers

Various coatings can be used to reduce or eliminate the emissions from underlying materials. These include laminates, veneers, and liquid-applied or dry powder coatings. Their effectiveness as barriers has been most extensively measured for reducing formaldehyde emissions and, to a lesser extent, controlling general VOC emissions. Emissions barriers for "hidden" surfaces and for exposed material edges can be particularly effective in controlling contaminant release.

### Material Conditioning and In-Place Curing

Airing new materials in a well-ventilated, clean space prior to installation in a building can be an effective means of reducing the typically high emissions that characterize new materials. Off-site opening of wrapped or tightly packaged materials to facilitate this "conditioning" phase is an important aspect of this simple strategy. Products that have been formulated to undergo a curing process that will result in reduced contaminant emissions can be sought during material selection. This is particularly relevant for products such as caulks, sealants, and adhesives that must be applied in wet form and for which effective in-place curing can have greatest benefit.

### Local Exhaust of Unavoidable Emissions

Where contaminant-generating equipment or activities can be localized in a specific area (such as with certain types of office equipment, kitchen/cafeteria operations, etc.), an effective strategy to limit the impact on IAQ is to provide local exhaust systems. Also see Strategy 6.2 – Provide Local Capture and Exhaust for Point Sources of Contaminants.

### Staged Entry of Materials

To the extent practical, materials that by their nature are highly absorptive (or "fleecy") need to be installed after completion of construction activities that release high levels of VOC contaminants (e.g., painting/staining and application of caulks, adhesives, and sealants). These absorptive materials include textiles, carpets/underlayments, acoustical ceiling tiles, open-plan office partition panels, and insulation materials. Without careful staging of their installation and storing prior to their installation, they can act as contaminant reservoirs (sinks), leading to long-term re-emission into the indoor air.

### Delayed Occupancy

It is important to delay building occupancy until a reasonable flush-out operation to reduce contaminant levels from early-phase product emissions has been completed. Building bake-out, in which the interior space is heated to between 95°F and 102°F (35°C and 39°C) in an attempt to speed emissions from materials and finishes, is discouraged because the effect has been shown to be temporary, tends to merely redistribute contaminant sources, and may damage building materials.

## Strategy 5.2

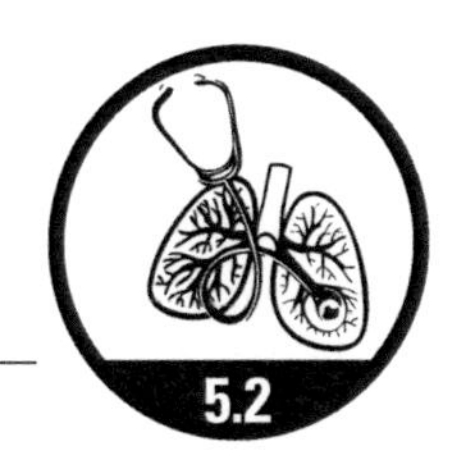
5.2

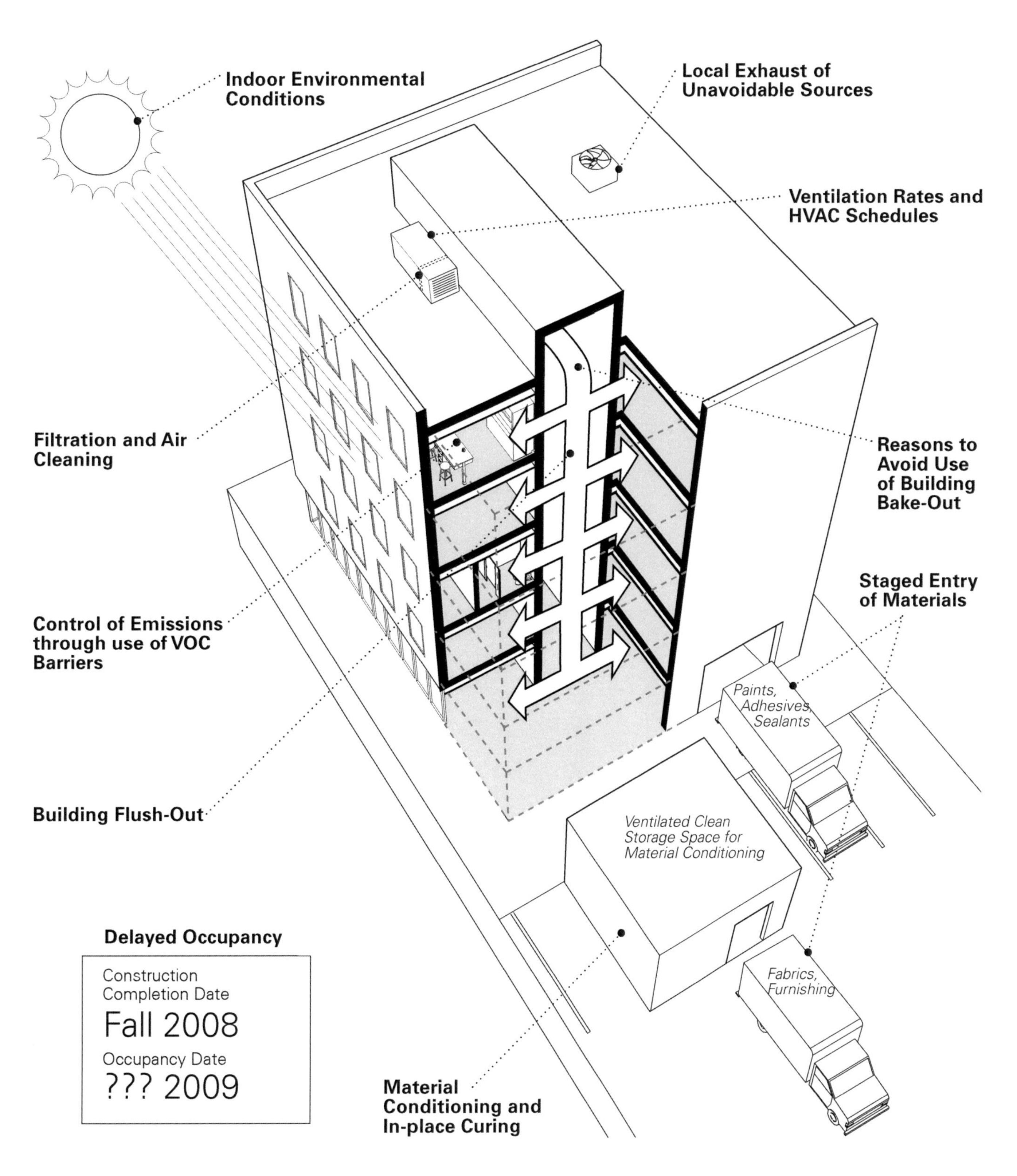
Indoor Environmental Conditions
Local Exhaust of Unavoidable Sources
Ventilation Rates and HVAC Schedules
Filtration and Air Cleaning
Reasons to Avoid Use of Building Bake-Out
Staged Entry of Materials
Control of Emissions through use of VOC Barriers
Paints, Adhesives, Sealants
Building Flush-Out
Ventilated Clean Storage Space for Material Conditioning
Delayed Occupancy
Construction Completion Date
Fall 2008
Occupancy Date
??? 2009
Fabrics, Furnishing
Material Conditioning and In-place Curing

**Building Flush-Out**

At completion of a new building, contaminant emissions from building materials and interior surfaces are typically at their highest. It is useful, therefore, to operate the building HVAC systems at a higher than normal ventilation rate for a period of time to help flush the building of these contaminants prior to occupancy and even during initial occupancy. The specific flush-out procedure employed must be adapted to local climatic/seasonal conditions.

**Ventilation Rates and HVAC Schedules**

Guidance and training to operation personnel (see Strategy 1.5 – Facilitate Effective Operation and Maintenance for IAQ) related to HVAC operation schedules needs to also consider provision of adequate ventilation during cleaning activities (see Strategy 5.3 – Minimize IAQ Impacts Associated with Cleaning and Maintenance) and any other activity where high emissions might be expected (e.g., painting, caulking, applying adhesives).

**Indoor Environmental Conditions**

In addition to considerations for areas that need to be able to withstand repeated wettings (Strategy 2.5 – Select Suitable Materials, Equipment, and Assemblies for Unavoidably Wet Areas), which can affect the emissions from building materials, temperature-induced increases in emissions can result from solar gains or from the use of radiant flooring. Special selection of materials is required in such situations, while general management of building temperature and humidity remains an assumed element of material emissions control.

**Filtration and Air Cleaning**

Gas-phase or particle-phase filtration can, if applied to HVAC system return airstreams, assist with the removal of IAQ contaminants. Ozone filtration of outdoor air can reduce the formation of ultrafine particulates and gaseous irritants formed through reaction with IAQ contaminants. Strategy 7.5 – Provide Particle Filtration and Gas-Phase Air Cleaning Consistent with Project IAQ Objectives provides additional guidance on this method to limit the impact of indoor material emissions.

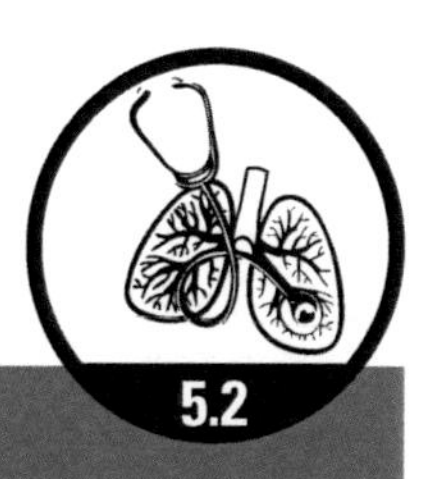


## Impact of Flush-Out on Similar Floors in New Office Building

As part of the Cx process for a new office building, the building's AHUs, toilet exhaust, and outdoor air systems were specified to run continuously during the last month of construction and for the first few weeks of occupancy. Air sampling conducted during this flush-out period revealed a significant difference in the VOC levels on two similarly designed and furnished floors (Figure 5.2-A). An investigation of the cause of this unexpected result revealed a malfunction in the outdoor air system for Floor A. Air change measurements revealed that while Floor B was ventilated at 0.76 ach, the ventilation rate of Floor A was essentially zero. Comparing VOC levels, it was found that the levels on Floor B were 64% to 92% lower than the corresponding levels on Floor A. This gave a clear indication of the effectiveness of flush-out in reducing contaminant levels in a newly furnished office space.

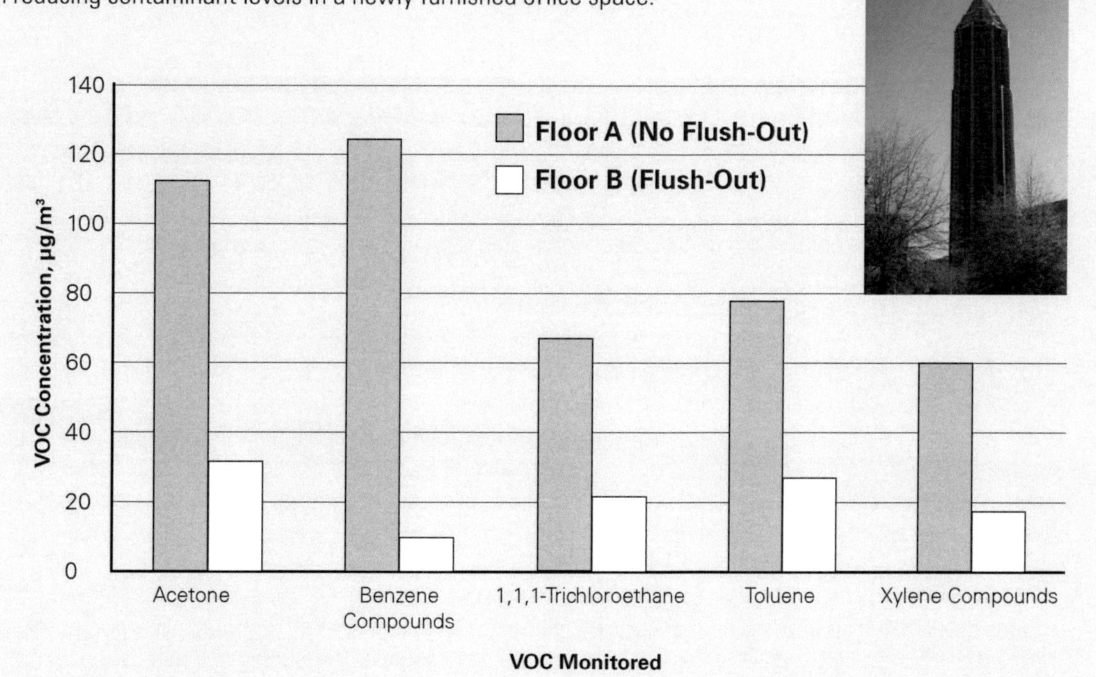


**Figure 5.2-A** Indoor VOC Concentrations on Two Typical Office Floors: Impact of Flush-Out
*Data source: Milam (1993). Inset photograph courtesy of H.E. Burroughs.*

# Minimize IAQ Impacts Associated with Cleaning and Maintenance



A clean indoor environment is generally considered an essential requisite for good IAQ. There is growing recognition, however, that if chosen poorly, cleaning agents and/or practices can have detrimental effects on indoor air. Designs that minimize the need for cleaning and that facilitate cleaning with non-toxic and non-corrosive agents and that provide good O&M documentation for proper cleaning methods can go a long way toward ensuring a clean and healthy indoor environment.

A preferred initial strategy is to prevent dirt from entering the building in the first place, thus reducing the need for cleaning. This strategy pays additional dividends by lowering operating costs (cleaning is expensive) and can be achieved through effective design of building approaches and dirt track-off systems (see Strategy 3.5 – Provide Effective Track-Off Systems at Entrances) and through improved HVAC filtration system design and maintenance (see Strategy 7.5 – Provide Particle Filtration and Gas-Phase Air Cleaning Consistent with Project IAQ Objectives).

**Select Durable Materials and Finishes that are Simple to Clean and Maintain**

As part of the design of the building, selecting interior materials and finishes that have surfaces that can be easily cleaned without strong chemical agents is also an important aspect of the overall strategy for controlling cleaning-associated IAQ impacts. Ensuring that these surfaces are also durable will reduce maintenance requirements as well as the IAQ impacts associated with replacement or refinishing. Particular attention should be paid to appropriate selection of flooring materials and surfaces used in restrooms.

**Recommend Cleaning Products with Minimal Emissions**

Cleaning will be an ongoing and essential requirement of building operation. This is why cleaning protocols and guidance need to be included in the O&M documentation (see Strategy 1.5 – Facilitate Effective Operation and Maintenance for IAQ). Cleaning products are diverse in function and chemical composition and may contain ingredients that are irritating or harmful to building occupants and cleaning staff. Formation of highly irritating secondary byproducts in the presence of indoor oxidants has been associated with certain cleaning products. Careful selection of cleaning products to avoid these potential problems is thus an important component of the O&M documentation.

**Provide Appropriate Storage for Cleaning Products**

Storage and handling of cleaning products in well-designed and ventilated janitorial closets is an important aspect of an overall strategy to minimize IAQ impacts associated with cleaning and maintenance. Provision of hot water taps and adequate mop sinks, proper dispensing systems for stock cleaning agents, moisture-resistant flooring materials, posted instructions for preparation of cleaning agents, and protocols in well-located closets will assist in the delivery of improved cleaning services.

**Recommend Cleaning Protocols that Will Have Minimal IAQ Impact**

Protocols for effective cleaning need to be included in O&M documentation and training. These ought to include the equipment employed in cleaning operations, the timing of cleaning activities, and provision of adequate building ventilation during and immediately following cleaning operations as well as the effective training of cleaning personnel in all these issues.

# Strategy 5.3

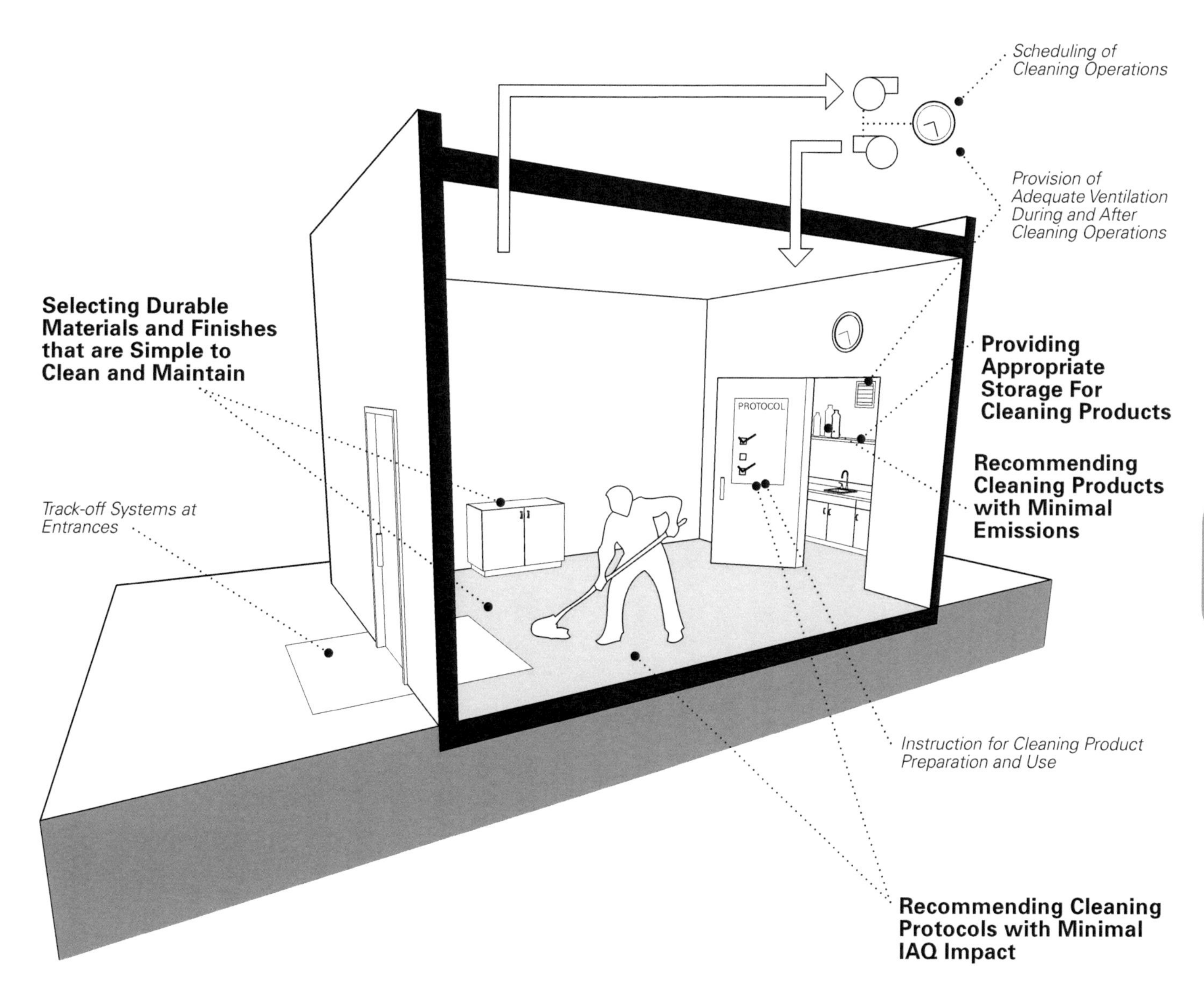


STRATEGY OBJECTIVE 5.3

## "Green Housekeeping" Program at Brooklyn Public Library

- *Reduction of Toxins.* Over 16 hazardous substances have been eliminated from Brooklyn Public Library's cleaning operations (including butoxyethanol, diacetone alcohol, dipropylene glycol, petroleum distillates, ethanolamine, ethyl ether, isobutane, isopropanol, methyl ether, naptha, and nonyl phenolethoxylate).

- *Reduction of Cleaning Products.* Staff estimates a reduction of approximately 50% in the amount of cleaning products used, primarily the result of using a proportioning chemical dispenser, which premixes cleaners and disinfectants for accurate dilution.

- *Packaging Waste Reduction.* Staff eliminated the use of 55 gal storage drums, which, in addition to being bulky and wasteful, were difficult and dangerous to handle.

- *Improved Efficiency.* The use of the proportioning dispenser was shown to save time. Staff believe the Green Housekeeping initiative boosted the morale of the custodial staff, increasing productivity as a result.

*Data source: NY (1999).*

5.3

# Capture and Exhaust Contaminants from Building Equipment and Activities

Building equipment and activities can be significant sources of indoor air contaminants. Among these are combustion products from fuel-burning equipment; exhaust from vehicles in enclosed parking garages; hazardous air pollutants from dry cleaners and nail and hair salons; particles and fumes from school laboratories, shops, and art classrooms; VOCs and ozone from office equipment; infectious agents from medical and dental procedure rooms; and odors from various sources. The Strategies discussed in this Objective can reduce the likelihood that these emissions will degrade IAQ.

- Combustion produces moisture, $CO_2$, oxides of nitrogen and sulfur, soot, and potentially CO. Strategy 6.1 – Properly Vent Combustion Equipment describes venting and combustion air requirements to limit occupant exposure to combustion products.
- Point sources such as large copiers and printers; nail care stations; certain workstations in laboratory, shop, and art classrooms; and commercial cooking equipment can produce contaminants that may cause irritation or illness. Strategy 6.2 – Provide Local Capture and Exhaust for Point Sources of Contaminants describes techniques to reduce users' and other occupants' exposure through well-designed exhaust and depressurization of the source area.
- Contaminants in exhaust air can re-enter the occupied space if exhaust ductwork is not well sealed, especially if the exhaust duct static pressure is higher than that in the surrounding area. Exhaust discharge can also be re-entrained into outdoor air intakes or windows. Strategy 6.3 – Design Exhaust Systems to Prevent Leakage of Exhaust Air into Occupied Spaces or Air Distribution Systems addresses duct sealing, fan location, and discharge design to reduce the risk of re-introducing exhaust to the occupied space.
- Many contaminant sources are too diffuse to be exhausted at the point of generation. Strategy 6.4 – Maintain Proper Pressure Relationships Between Spaces describes methods to control contaminant transfer from such spaces as enclosed parking garages, natatoriums, dry cleaning shops, hair salons, and bars through space layout and compartmentalization and control of space-to-space pressures.

Other Strategies that can affect or be affected by exhaust and space depressurization include the following:

- Strategy 1.3 – Select HVAC Systems to Improve IAQ and Reduce the Energy Impacts of Ventilation
- Strategy 2.2 – Limit Condensation of Water Vapor within the Building Envelope and on Interior Surfaces
- Strategy 2.3 – Maintain Proper Building Pressurization
- Strategy 3.2 – Locate Outdoor Air Intakes to Minimize Introduction of Contaminants
- Strategy 3.3 – Control Entry of Radon
- Strategy 3.4 – Control Intrusion of Vapors from Subsurface Contaminants

# Objective 6

**6.1** Properly Vent Combustion Equipment

**6.2** Provide Local Capture and Exhaust for Point Sources of Contaminants

**6.3** Design Exhaust Systems to Prevent Leakage of Exhaust Air into Occupied Spaces or Air Distribution Systems

**6.4** Maintain Proper Pressure Relationships Between Spaces

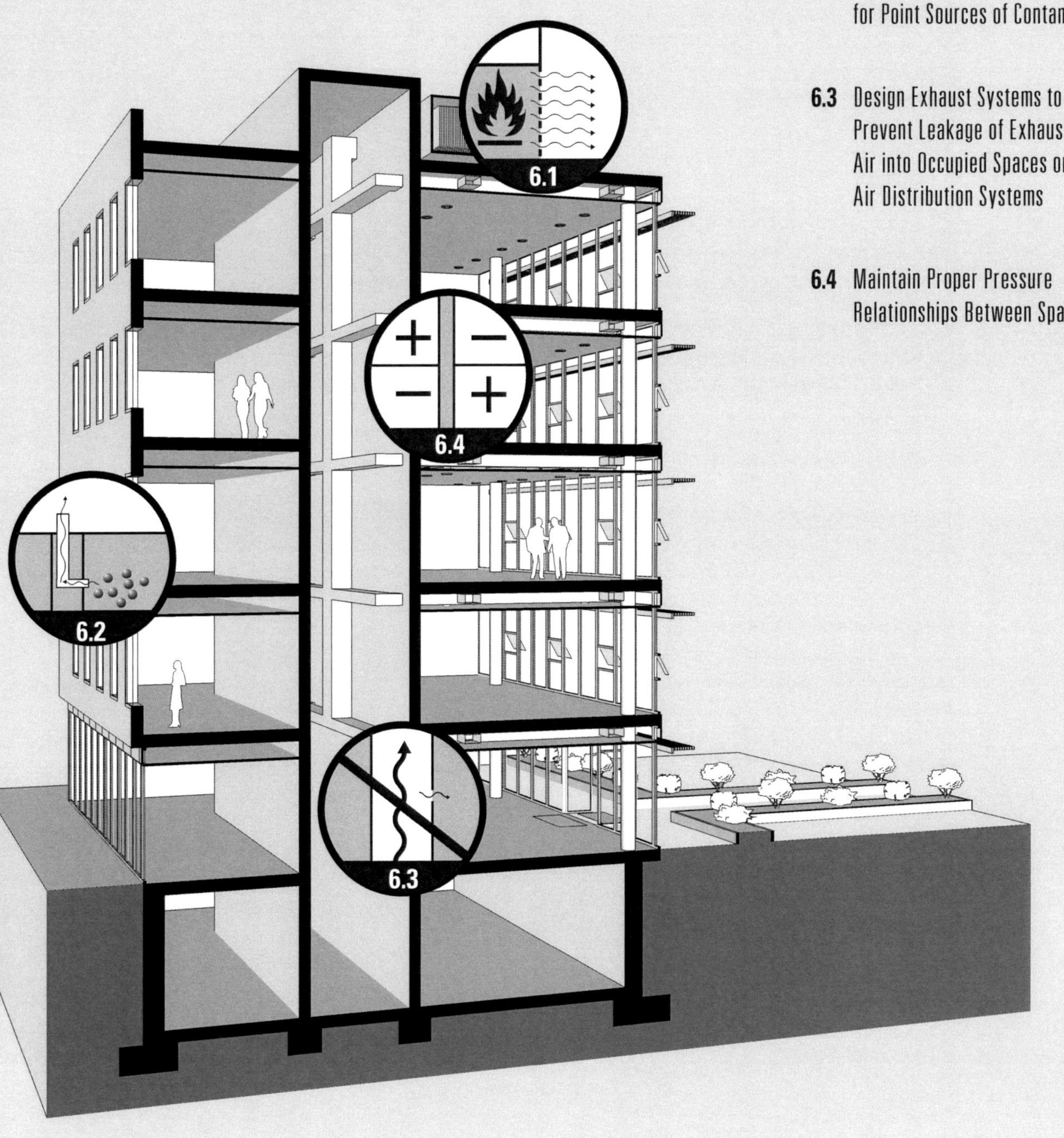

# Properly Vent Combustion Equipment



Many types of combustion equipment and appliances are used in buildings. Since combustion produces harmful byproducts (e.g., CO, $NO_2$, and fine particles), it is important to control the flow of these byproducts through carefully designed venting and exhaust systems and through provisions for supplying outdoor air for combustion.

**Capture and Exhaust of Combustion Byproducts**

The type of exhaust capture system used will depend on the fuel, process, and type of equipment being vented.

- Chimney (natural draft) systems rely on the buoyancy of the warm combustion products in the chimney or stack (relative to the cooler and denser surrounding air) to produce a natural draft that exhausts the combustion products from the building.
- Induced draft systems use a fan on the downstream side of the combustion chamber to pull combustion products through the combustion chamber and exhaust them from the building.
- Forced draft systems use a fan on the upstream side of the combustion chamber to push air through the combustion chamber and exhaust combustion products from the building.

Regardless of the type of system used, all components of any capture and exhaust system must be selected to properly function under the expected operating conditions, including the temperature and other properties of the exhaust air. The system must then be designed and installed to effectively remove the products of combustion.

As important as the size and operation of the exhaust system to remove combustion products is the availability of an adequate supply of makeup air for combustion. An inadequate amount of makeup air may lead to incomplete combustion, resulting in an increase of the harmful combustion byproducts, particularly CO. The inadequate supply of makeup air can also result in negative pressures at the equipment burner, which can cause back-drafting, where the exhaust gasses are pulled back down through the exhaust vents.

**Operation and Maintenance**

There is a greater potential for exposure to harmful combustion products if the combustion equipment itself is not well maintained. Installation of combustion equipment must therefore provide adequate access for proper maintenance according to the manufacturer's instructions. It is also recommended that monitoring of the design and installation of combustion equipment and exhaust and supply duct systems be included as part of the building Cx process. The O&M manual should provide information on the recommended frequency of inspections to ensure that combustion equipment is operating properly.

**Commissioning**

Given the potential hazards from combustion equipment, the design and installation of combustion equipment, along with the exhaust and supply duct systems, should be included as a part of the building Cx process.

# Strategy 6.1

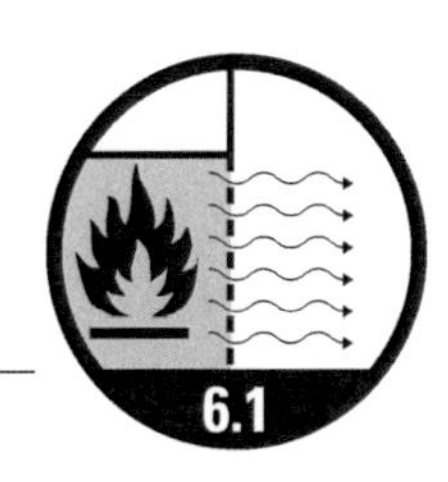


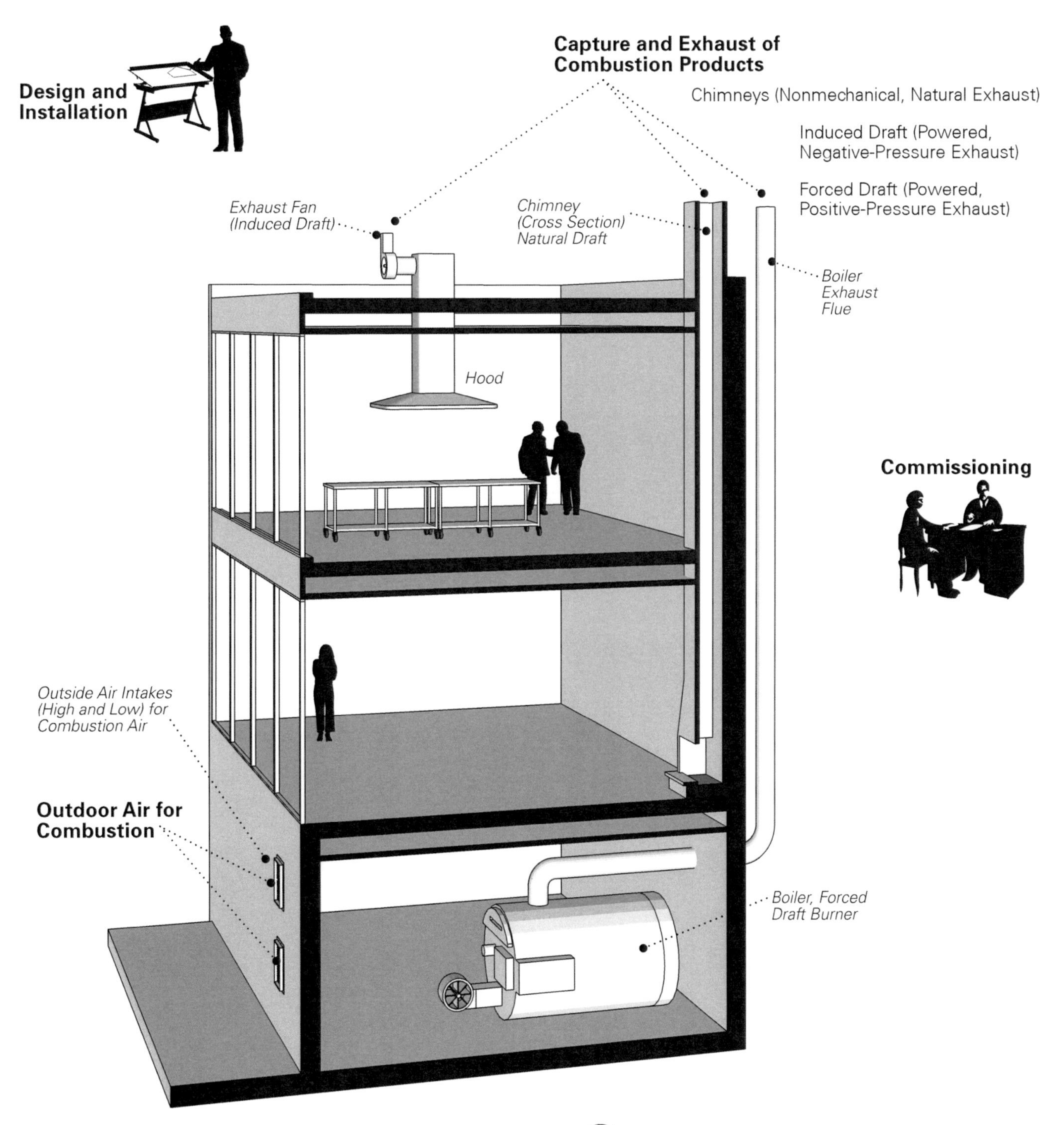

STRATEGY OBJECTIVE 6.1

# Provide Local Capture and Exhaust for Point Sources of Contaminants



In ASHRAE Standard 62.1, a contaminant is defined as "an unwanted airborne constituent that may reduce acceptability of the air" (ASHRAE 2007a, p. 4). By this definition, in the indoor environment today, contaminants are generated as a part of such processes as cooking, commercial laundries, scientific procedures and experimentation, generation and reproduction of paper materials, personal nail treatments, and woodworking and metal shop procedures as well as in areas where chemicals may be utilized extensively (such as natatoriums, photographic material facilities, and hair salons). The potential impacts on the occupants in these spaces and the surrounding areas include skin irritations, nose/sinus irritations, objectionable odors, and damage to interior building construction materials and/or finishes. The effective local capture and exhaust for point sources of contaminants can significantly reduce the impact of these contaminants on the occupants in the area in which the contaminant is generated and surrounding occupied spaces.

To be effective, the exhaust system design needs to achieve the following:

- Capture the exhaust as close to the source as possible, and exhaust directly to the outdoors.
- Maintain the area in which these contaminants are generated at a negative pressure relative to the surrounding spaces to reduce the potential impact on occupants in adjacent spaces.
- Enclose and exhaust the areas where contaminants are generated.

Capturing contaminants as close to the source as possible, as with an exhaust hood, significantly increases the capture rate of contaminants and reduces the exposure of occupants to these contaminants.

Exhausting directly to the outdoors removes the contaminants from the building. The location and height of the exhaust discharge outside the building is important to prevent re-entrainment of the contaminants into the building or surrounding buildings.

Maintaining the area in which these contaminants are generated at a negative pressure relative to the surrounding space reduces the potential migration of contaminated air into adjacent occupied spaces.

Enclosing the area where the contaminants are being exhausted assists in maintaining the space under negative pressure and also adds a physical barrier to the potential migration of contaminants to adjacent spaces.

The principles of capturing contaminants close to the source and exhausting to the outdoors, maintaining the area under negative pressure, and enclosing the area are part of a contaminant control system and are meant to be used in concert with one another and not as substitutes for each other.

# Strategy 6.2

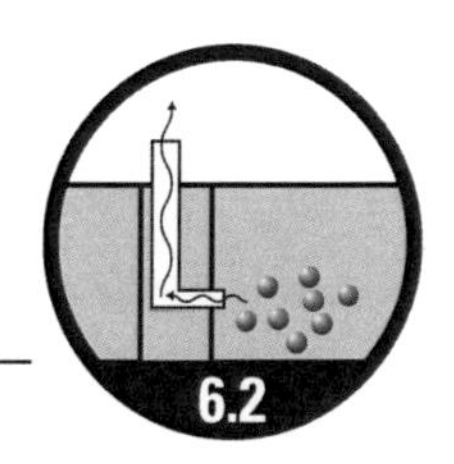


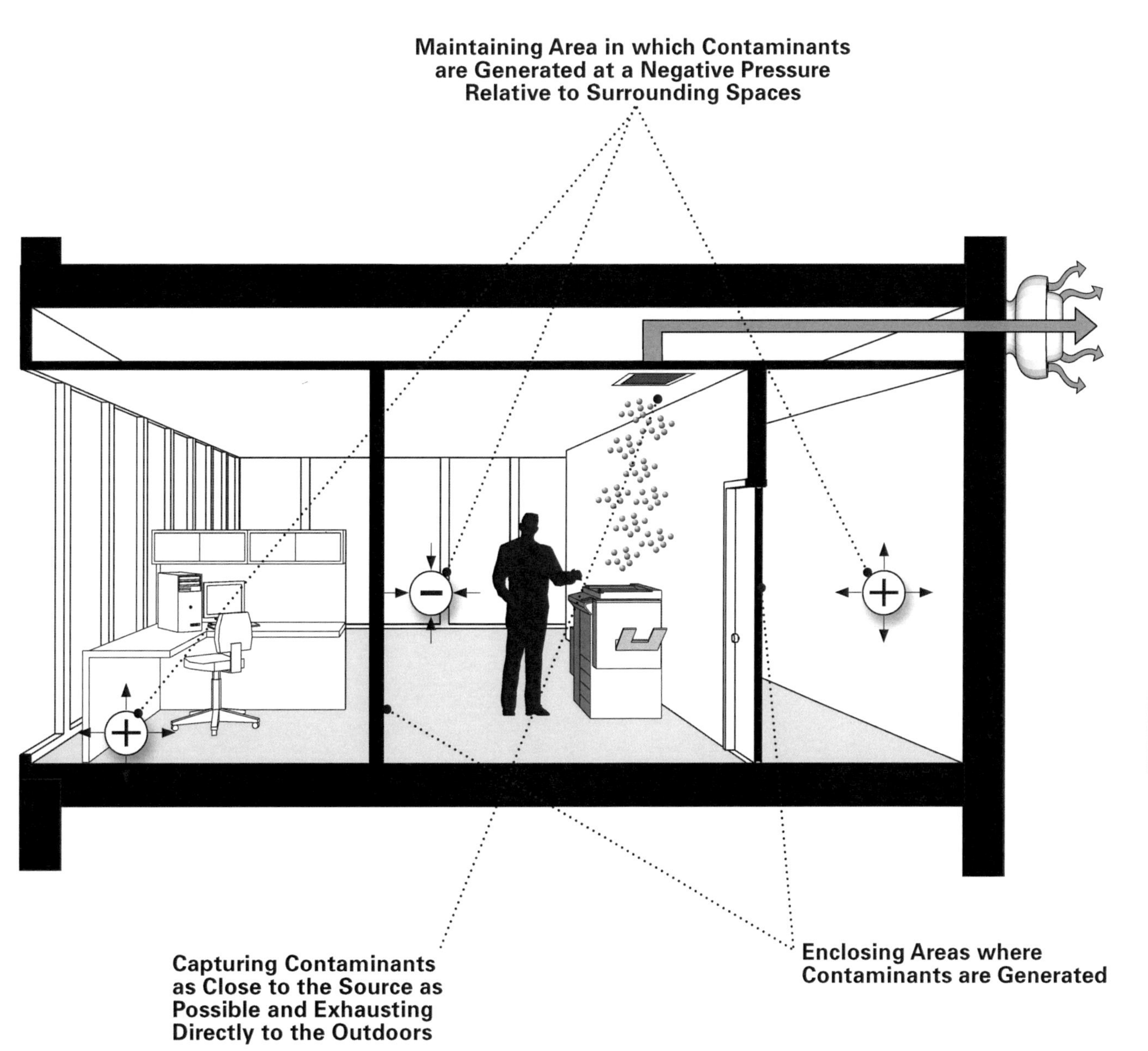


STRATEGY OBJECTIVE 6.2

## Lack of Exhaust—Indoor Swimming Pool

**Figure 6.2-A** Corridor to Indoor Swimming Pool
*Photograph courtesy of H.E. Burroughs.*

Figure 6.2-A shows a hotel facility in the southeastern United States. The photograph was taken from a corridor connecting the main hotel building to the remote pod where the indoor swimming pool is located. The swimming pool area is not provided with an exhaust fan, which results in several IAQ issues:

1. chemicals utilized for the pool treatment are not exhausted from the space,
2. the lack of exhaust in the space contributes to condensation on the fenestration in the space, and
3. the adjacent space (main hotel building) is impacted by the chemical contaminants and odors from the pool area, which are evident a substantial distance into the hotel building wing. Moisture from the pool is also likely to migrate into the hotel building. The migration of odors and/or moisture in this case study is encouraged by the main hotel building, and not the indoor swimming pool pod, operating under a negative pressure relative to surrounding spaces and the outdoor environment.

STRATEGY OBJECTIVE 6.2

# Design Exhaust Systems to Prevent Leakage of Exhaust Air into Occupied Spaces or Air Distribution Systems



Exhaust systems are required to remove odors and contaminants from the indoor environment. Areas that require exhaust include toilet rooms, soiled laundry storage rooms, pet shops (animal areas), areas where chemicals may be utilized extensively (such as natatoriums, photographic material facilities, and hair salons), and spaces where contaminants are generated as a part of such processes as cooking, scientific procedures and experimentation, generation and reproduction of paper materials, personal nail treatments, and woodworking and metal shop procedures. The potential impacts on the occupants in these spaces and the surrounding areas include skin irritations, nose/sinus irritations, objectionable odors, and damage to interior building construction materials and/or finishes. In addition, if the building design requires the use of a smoke control system, the potential impact of leakage of the exhaust system can have health or life safety consequences.

In order to reduce the potential impact on the occupants in spaces outside of the area of the odor or contaminant, air must be effectively exhausted directly from the space. This topic is covered in Strategy 6.2 – Provide Local Capture and Exhaust for Point Sources of Contaminants.

In addition to effectively exhausting the area of odors or contaminants, it is critical that the exhaust air be conveyed to the outdoors and discharged in a manner to reduce leakage of the exhaust air into surrounding occupied spaces or into air distribution systems.

**Effectively Seal Ductwork to Limit Leakage from the Duct System**
Sealing of exhaust ductwork is paramount in order to reduce the potential for leakage of contaminants into plenum spaces or adjacent areas. Leakage rates vary significantly based on the method of duct fabrication, the assembly methods, and the quality of workmanship in the installation. There are various classes of sealing recognized by ASHRAE (ASHRAE 2009) and SMACNA (SMACNA 2005), depending on the specific application.

**Provide Outdoor Discharge Position and Configuration to Limit Entrainment of Exhaust Air into Fresh Air Ventilation Systems or Adjacent Buildings**
The location of the exhaust discharge and configuration is a critical component in reducing the potential for an exhaust system to have an impact on outdoor air ventilation systems and/or adjacent facilities or buildings. The analysis must not only consider the exhaust system composition, equipment, and installation but also take into account the effects of architectural screen walls/enclosures, prevailing wind patterns, and height and proximity of adjacent buildings.

**Maintain Exhaust Ducts in Plenum Spaces under a Negative Pressure**
To assist in reducing the potential impact of exhaust systems on adjacent spaces, the duct must be maintained at a negative pressure as it passes through any plenum spaces. This is a specific requirement in the *International Mechanical Code* (*IMC*; ICC 2006a) and results in defining the potential location for fans utilized with the exhaust system.

# Strategy 6.3

6.3

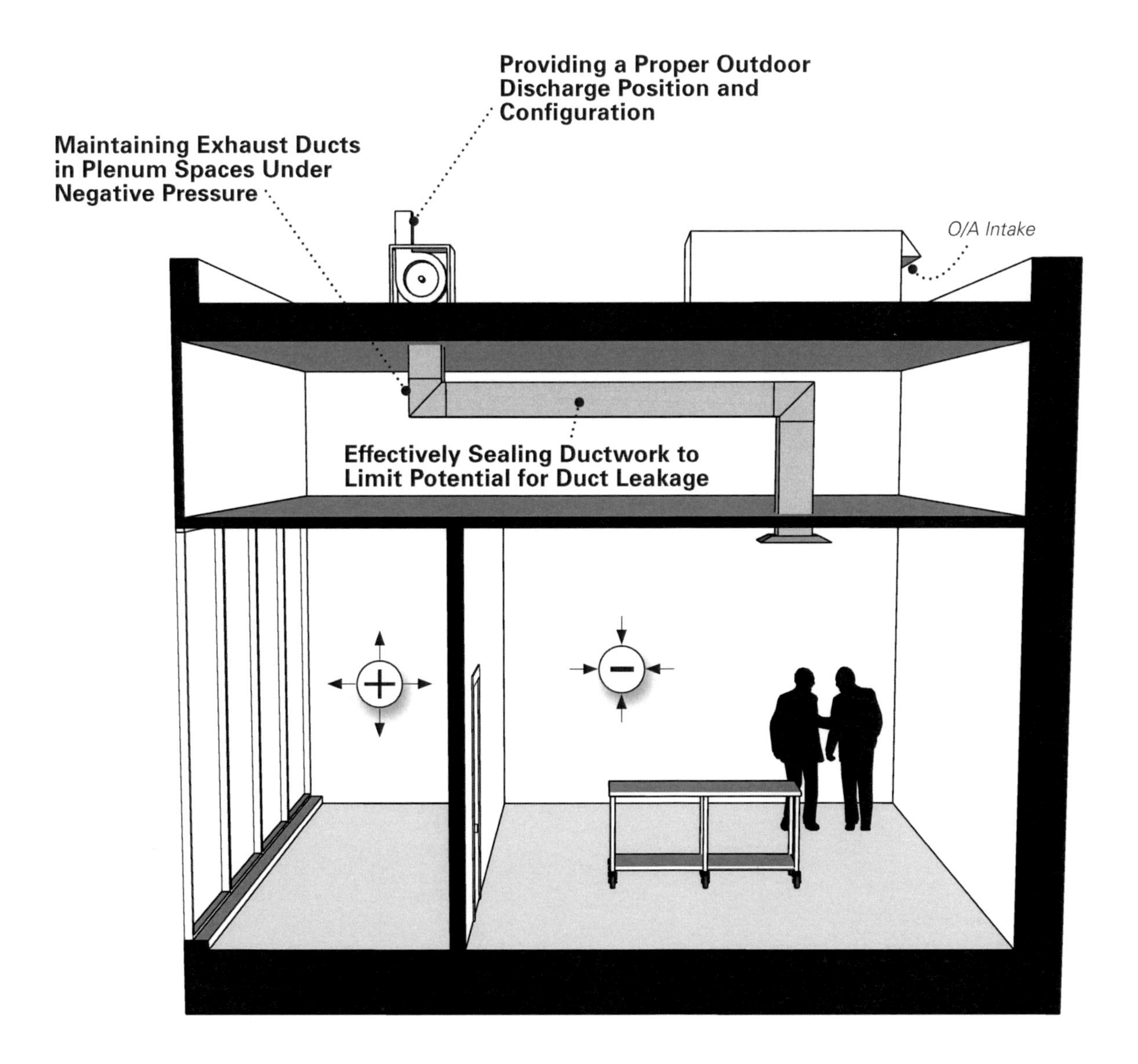
Providing a Proper Outdoor Discharge Position and Configuration
Maintaining Exhaust Ducts in Plenum Spaces Under Negative Pressure
O/A Intake
Effectively Sealing Ductwork to Limit Potential for Duct Leakage

## Improper Separation between Exhaust and Intake Airstreams

**Figure 6.3-A** Location of Exhaust Fan

**Figure 6.3-B** Relative Location of Exhaust Fan to Intake Louvers

Figures 6.3-A and 6.3-B show an office building in California and reflect the roof installation of a mechanical exhaust system (toilet exhaust fan shown with dome on top) in close proximity to and upwind of the air intake location for one of the largest air-handling systems for the building. The problem was remedied, post-construction, by installing a sheet metal surround to raise the height of the air discharge above the height of the intake louver, allowing the exhaust plume to dissipate without entering the air intake.

*Photographs courtesy of Hal Levin.*

6.3

# Maintain Proper Pressure Relationships Between Spaces



Proper space pressurization reduces moisture and contaminant transfer between adjacent spaces, thereby reducing contamination of occupied spaces and unwanted condensation and mold growth. Space pressurization refers to the static pressure difference between the adjacent spaces of a building, with the air tending to move from higher-pressure spaces to lower-pressure spaces. This static pressure difference will influence where exfiltration and infiltration occur across the adjacent spaces. Maintaining proper pressure relationships between adjacent spaces is critical to ensure airflow in the preferred direction, from clean spaces to dirty spaces. Many HVAC systems are designed to achieve a space-to-space differential pressure from 0.01 to 0.05 in. w.c. (2.5 to 12.5 Pa) where pressure relationships are needed.

**Space Usage**
To determine the described pressure relationship between spaces, the usage of building spaces, along with their moisture and contaminant sources or conditions, need to be identified. Decisions then need to be made about which will be positively or negatively pressurized.

**Space Layout**
If moisture or contaminants are a concern, it is helpful to select a space layout within the building early in the design process for the most advantageous movement of air. For example, consider the locations of spaces in exterior zones versus interior zones. If a space is required to be negative relative to adjacent spaces, consider locating this space on an interior zone to reduce possible infiltration of unconditioned air into the space from outdoors.

**Space Envelope**
The effectiveness of the space pressurization is reduced by the leakiness of the space envelope. Therefore, the space envelope needs to be designed to limit exfiltration, infiltration, and leakage. Strategy 2.2 – Limit Condensation of Water Vapor within the Building Envelope and on Interior Surfaces provides information on the design and construction of the space envelope.

**Compartmentalization**
If space pressurization is not an option, sealing and other construction techniques can be used to compartmentalize spaces to contain contaminants and moisture.

**HVAC System**
A variety of issues need to be addressed to ensure appropriate space-to-space pressure control. These include design airflow rates, airflow measurements and monitoring control, negative relative pressures in return air plenums, and duct leakages that can compromise pressurization control and contaminate spaces with both pollutants and moisture.

**Verification**
A number of steps can be taken to ensure that desired space pressurization is attained. These include verifying proper construction of space envelopes, including design information on contract documents; performing testing and balancing to verify all airflows; and performing pressure differential mapping.

# Strategy 6.4

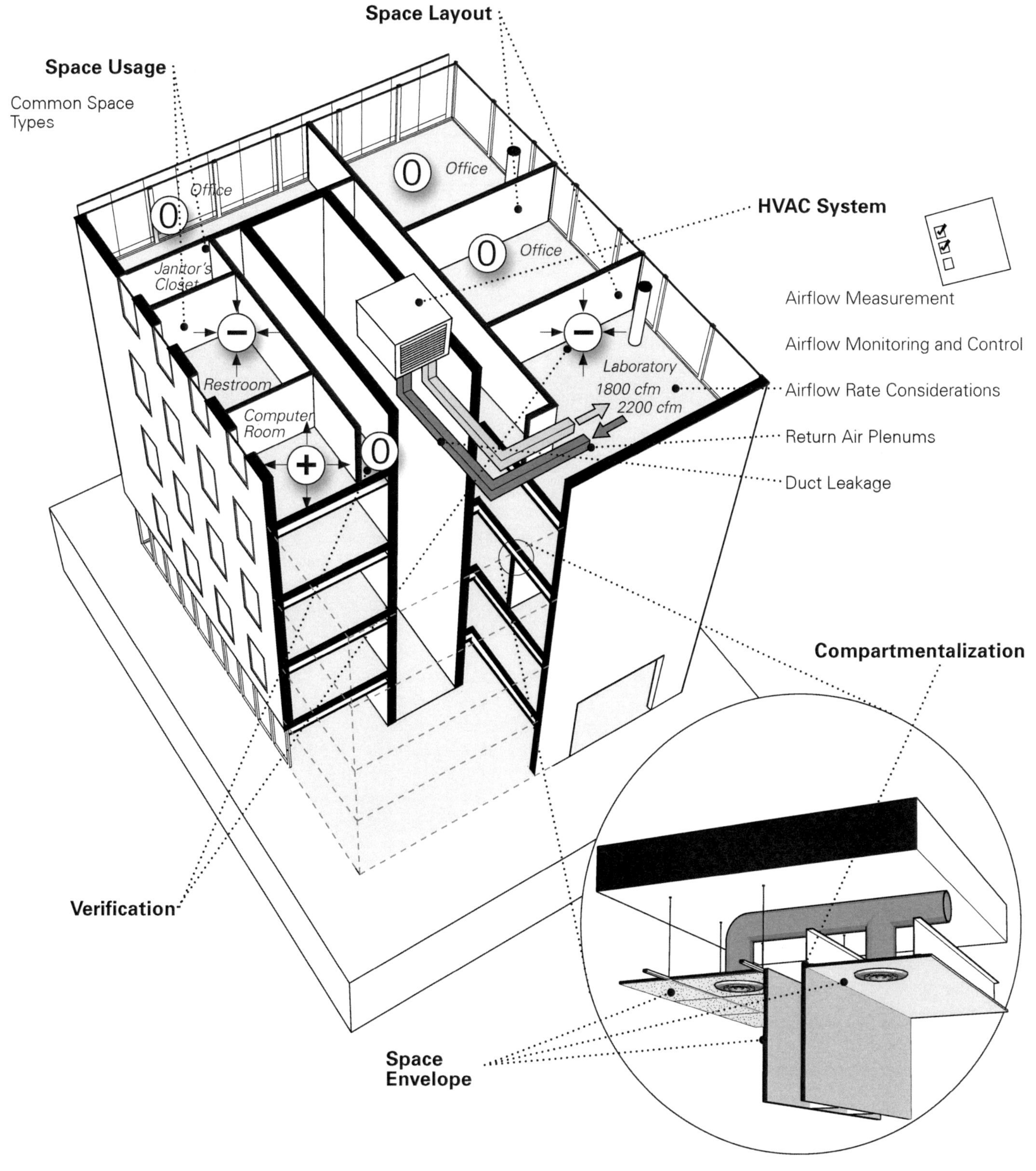

Space Layout
Space Usage
Common Space Types
Office
Office
Office
Janitor's Closet
Restroom
Computer Room
Laboratory
1800 cfm
2200 cfm
HVAC System
Airflow Measurement
Airflow Monitoring and Control
Airflow Rate Considerations
Return Air Plenums
Duct Leakage
Compartmentalization
Verification
Space Envelope

## Noxious Spa Odors Invade a Hotel

**Figure 6.4-A** Jacuzzi/Pool Area Emitting Odors

**Figure 6.4-B** Ill-Fitting Double Glass Doors

*Photographs courtesy of H. E. Burroughs.*

In this low-rise hotel facility, the guest exercise room and spa is located in a separate structure that also contains a large multi-person heated jacuzzi/pool (Figure 6.4-A). Access to the facility from the main guest room wing is through ill-fitting double glass doors at the end of the guest wing corridor (Figure 6.4-B). As guests and visitors to the facility approach the spa entrance, the odor of chlorine from the pool area is noticeable halfway down the building wing. It becomes increasingly noxious closer to the entrance doors. To users of the facility, as well as to guests in nearby rooms, it is obvious that the odors from the water treatment chemicals are migrating into the corridor and sleeping rooms, even though the spa facility is in a separate building pod.

The corridor leading to the spa area is continuous and open to the hotel lobby and breakfast/kitchen area. What appears to be happening is that the breakfast/kitchen area exhaust system is depressurizing the hotel corridor, making the corridor negatively pressurized relative to the spa. The spa odors are thus drawn into the corridor. In addition, individual room exhausts exacerbate the problem by inducing the contaminant from the corridor into individual rooms. An unseen and less obvious problem is that the same odorous airstream is also inducing steam and high humidity from the spa into the conditioned space of the guest rooms.

This demonstrates how important it is to be diligent to establish proper pressure relationships within the building so that air does not flow from contaminated or moisture-laden areas into clean occupied spaces. Specifically in this case, the design properly incorporated an exhaust system in the breakfast/kitchen area but did not account for its impact on the corridor and spa. Similar problems can occur in other types of buildings.

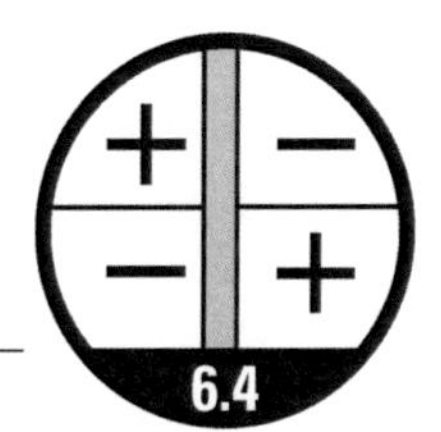

6.4

# Reduce Contaminant Concentrations through Ventilation, Filtration, and Air Cleaning

Design for IAQ should focus first on reducing contaminant sources and then on capturing and exhausting contaminants close to their source. Remaining contaminants should be diluted with ventilation air or reduced by filtration and gas-phase air cleaning (FAC). Inadequate ventilation increases the likelihood of adverse health effects and IAQ complaints. Insufficient FAC allows outdoor contaminants to be brought into the building, indoor contaminants to be recirculated, and dirt to accumulate in air-handling systems.

- Minimum ventilation requirements are described in Strategy 7.1 – Provide Appropriate Outdoor Air Quantities for Each Room or Zone.
- Poor control of minimum outdoor air delivery can waste energy if the flow is too high and degrade IAQ if the flow is too low. Strategy 7.2 – Continuously Monitor and Control Outdoor Air Delivery describes options to ensure that design airflows are actually delivered.
- To dilute contaminants effectively, ventilation air must be delivered to the breathing zone. The effectiveness of different systems in achieving this varies considerably, and failure to account for this can result in significant underventilation. Strategy 7.3 – Effectively Distribute Ventilation Air to the Breathing Zone describes procedures to calculate air distribution effectiveness and the impact of differences in effectiveness on energy use and costs.
- The percentage of outdoor air required by a system that serves multiple spaces can be difficult to determine and, for VAV systems, varies over time. Failure to properly account for these factors can result in poor ventilation. Strategy 7.4 – Effectively Distribute Ventilation Air to Multiple Spaces describes proper design of such systems.
- FAC can remove a substantial fraction of contaminants from incoming outdoor air, reduce recirculation of indoor contaminants, and reduce accumulation of dirt in air-handling systems. Strategy 7.5 – Provide Particle Filtration and Gas-Phase Air Cleaning Consistent with Project IAQ Objectives provides guidance on FAC selection.
- Occupant perception of IAQ is closely correlated to thermal comfort. Strategy 7.6 – Provide Comfort Conditions that Enhance Occupant Satisfaction addresses design for thermal comfort and integration of comfort and ventilation design.

Decisions made very early in the design phase may limit the project team's ability to provide good ventilation and FAC. These issues are discussed in the following Strategies:

- Strategy 1.1 – Integrate Design Approach and Solutions
- Strategy 1.3 – Select HVAC Systems to Improve IAQ and Reduce the Energy Impacts of Ventilation
- Strategy 3.2 – Locate Outdoor Air Intakes to Minimize Introduction of Contaminants

Strategies to reduce ventilation energy use are discussed in Objective 8 – Apply More Advanced Ventilation Approaches.

# Objective 7

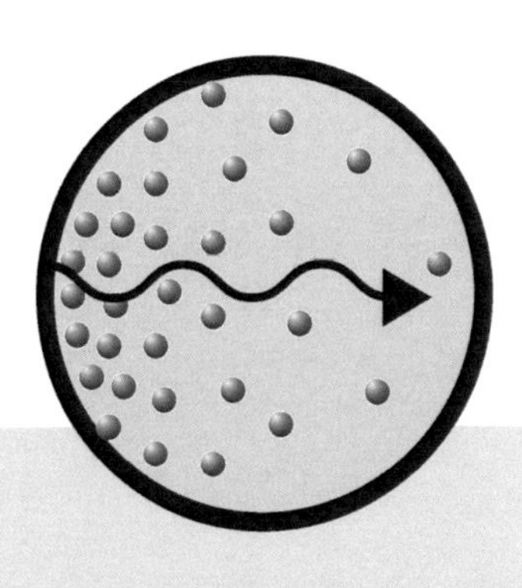

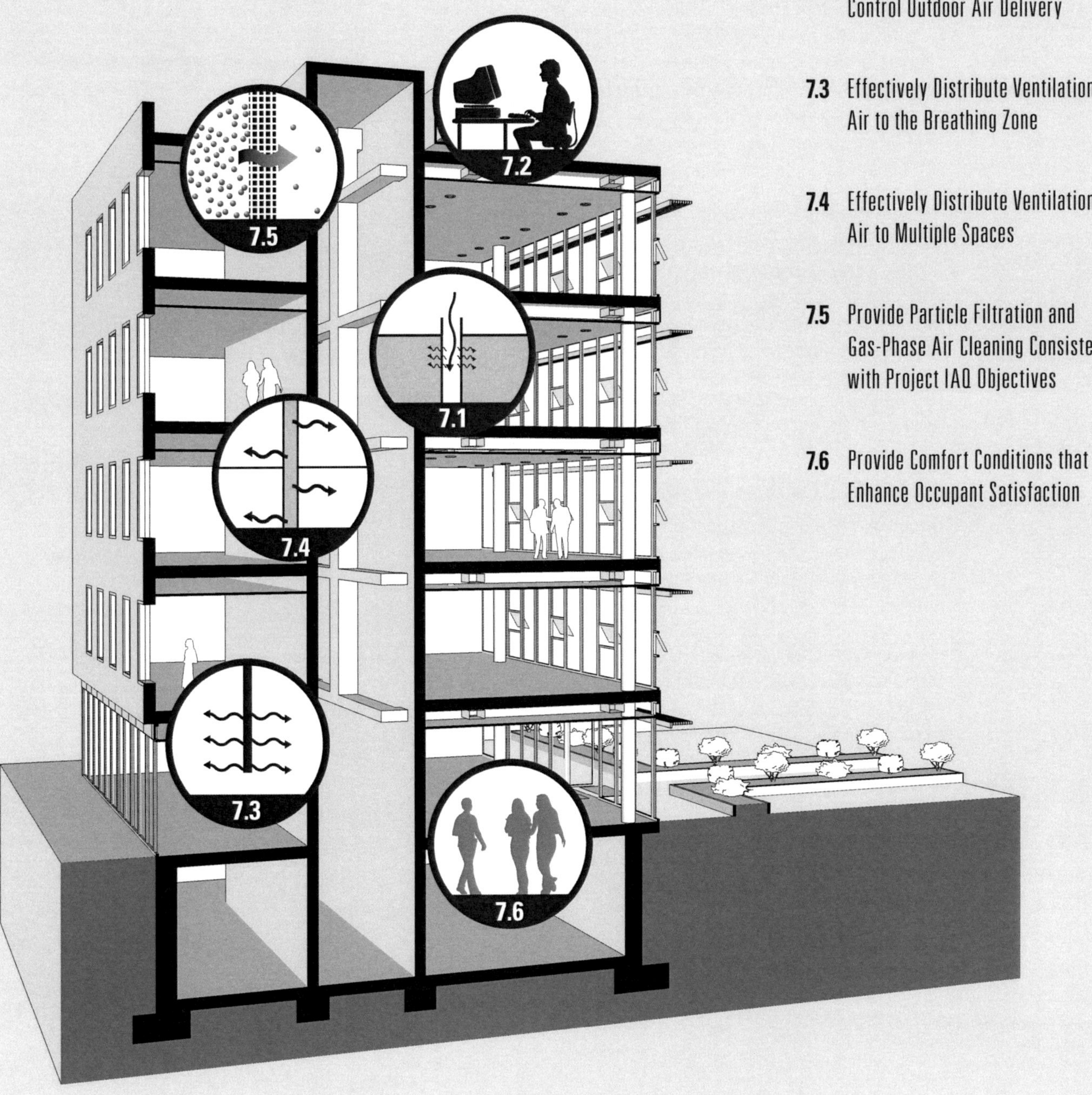


**7.1** Provide Appropriate Outdoor Air Quantities for Each Room or Zone

**7.2** Continuously Monitor and Control Outdoor Air Delivery

**7.3** Effectively Distribute Ventilation Air to the Breathing Zone

**7.4** Effectively Distribute Ventilation Air to Multiple Spaces

**7.5** Provide Particle Filtration and Gas-Phase Air Cleaning Consistent with Project IAQ Objectives

**7.6** Provide Comfort Conditions that Enhance Occupant Satisfaction

# Provide Appropriate Outdoor Air Quantities for Each Room or Zone



Outdoor air has been provided to indoor spaces for centuries, but the nature of building ventilation changed with the advent of electricity and the ability to provide ventilation to buildings mechanically, without relying on natural drafts. Ventilation with outdoor air is required for all occupied spaces. Inadequate outdoor air ventilation rates can result in poor IAQ and the potential for adverse health effects and reduced productivity for occupants, along with increased occupant complaints.

The Ventilation Rate Procedure in ASHRAE Standard 62.1-2007 (ASHRAE 2007a), specifies minimum ventilation rates for the U.S. Local building codes usually reference or include these rates but may differ in various ways from the standard. If local codes require more ventilation than specified in the standard, the local code requirements must be met. After the designer determines the outdoor air required for each zone, the quantity of air for the ventilation system must be adjusted to account for air distribution effectiveness and air-handling system ventilation efficiency.

ASHRAE Standard 62.1-2007 specifies two distinct ventilation rate requirements. The first is a per-person requirement to dilute pollutant sources associated with human activity that are considered to be proportional to the number of occupants. The second is a per-unit-area requirement designed to dilute pollutants generated by building materials, furnishings, and other sources not associated with the number of occupants.

The ventilation rates are specific to the type of occupant activity. For example, the outdoor air ventilation rates for different parts of an office building may vary depending on the occupant activity in the zones. Differences in occupant activity requiring different ventilation rates are evident in the graphical guide to the detailed information in Part II.

During short-term episodes of poor outdoor air quality, ventilation can be temporarily decreased using a short-term conditions procedure from ASHRAE Standard 62.1-2007. Similarly, consideration may be given to increasing outdoor air ventilation rates beyond those required in the standard where the quality of the outdoor air is high and the energy consumed in conditioning it is not excessive.

Information related to providing adequate outdoor air ventilation is also discussed in Strategy 7.3 – Effectively Distribute Ventilation Air to the Breathing Zone, Strategy 7.4 – Effectively Distribute Ventilation Air to Multiple Spaces, and Strategy 8.5 – Use the ASHRAE Standard 62.1 IAQ Procedure Where Appropriate.

# Strategy 7.1

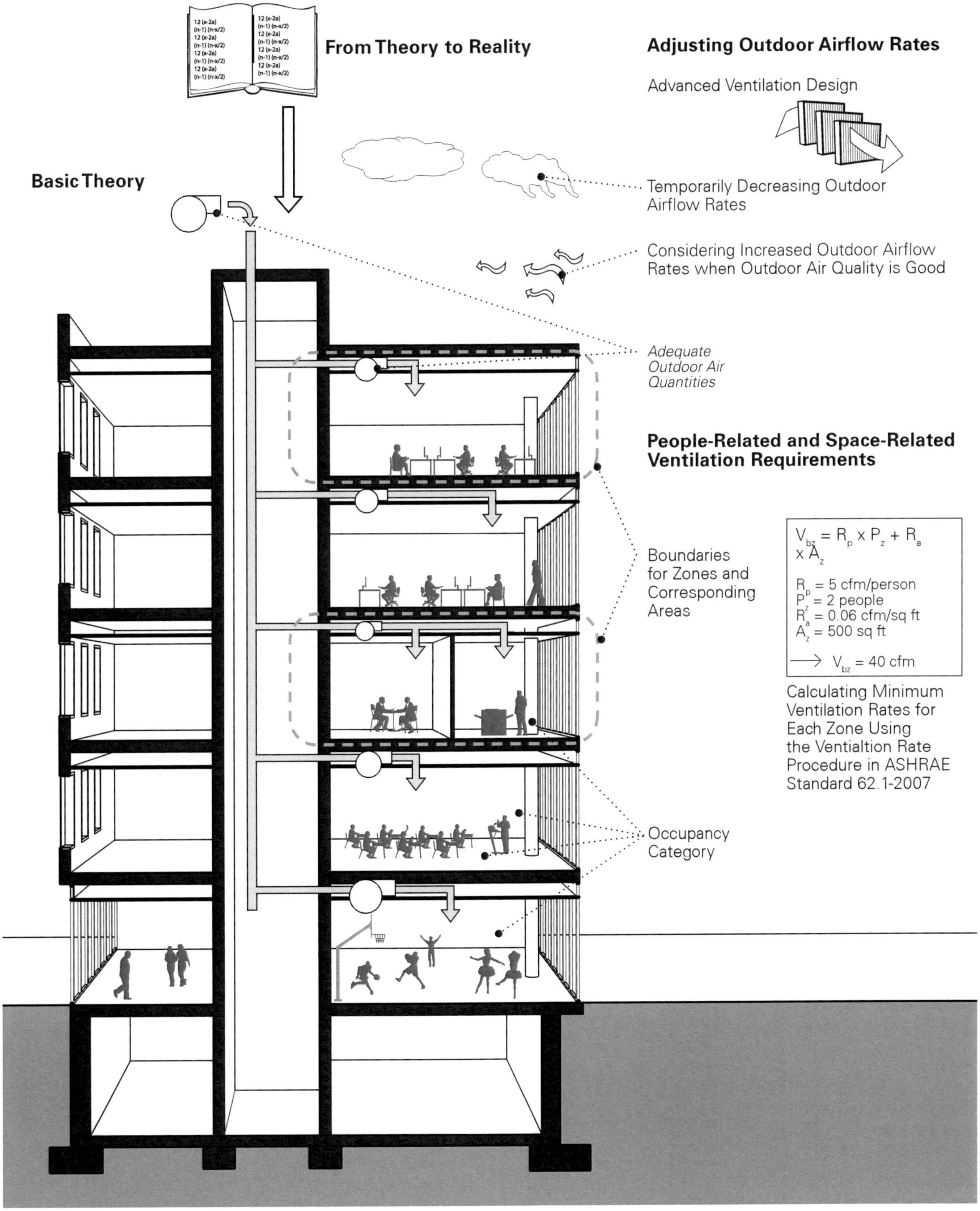

7.1

**From Theory to Reality**

**Basic Theory**

**Adjusting Outdoor Airflow Rates**

Advanced Ventilation Design

Temporarily Decreasing Outdoor Airflow Rates

Considering Increased Outdoor Airflow Rates when Outdoor Air Quality is Good

*Adequate Outdoor Air Quantities*

**People-Related and Space-Related Ventilation Requirements**

Boundaries for Zones and Corresponding Areas

$V_{bz} = R_p \times P_z + R_a \times A_z$

$R_p$ = 5 cfm/person
$P_z$ = 2 people
$R_a$ = 0.06 cfm/sq ft
$A_z$ = 500 sq ft

→ $V_{bz}$ = 40 cfm

Calculating Minimum Ventilation Rates for Each Zone Using the Ventialtion Rate Procedure in ASHRAE Standard 62.1-2007

Occupancy Category

## Ventilation and Performance

**Figure 7.1-A** Measuring Student Performance

Two recent studies have documented the associations between ventilation and student attendance and classroom performance. The first study (Shendell et al. 2004) documented the associations between classroom attendance in Washington and Idaho and $CO_2$ concentrations, which were used as a surrogate for ventilation rates. For classrooms where the difference between indoor and outdoor $CO_2$ concentrations exceeded 1000 ppm (1800 mg/m$^3$), student absences were 10%–20% higher than for classrooms where the difference in $CO_2$ was below 1000 ppm (1800 mg/m$^3$).

A second study (Wargocki and Wyon 2006) examined academic performance in a controlled classroom situation in Denmark where ventilation and temperature were varied. The authors reported that "increasing the outdoor air supply rate and reducing moderately elevated classroom temperatures significantly improved the performance of many tasks (Figure 7.1-A), mainly in terms of how quickly each pupil worked (speed) but also for some tasks in terms of how many errors were committed (% errors, the percentage of responses that were errors). The improvement was statistically significant at the level of $P \leq 0.05$" (p. 26).

These and other studies, including those conducted in office environments, are summarized in the IAQ Scientific Findings Resource Bank (IAQ-SFRB) (http://eetd.lbl.gov/ied/sfrb/sfrb.html).

7.1

STRATEGY OBJECTIVE 7.1

# Continuously Monitor and Control Outdoor Air Delivery



Accurate monitoring and control of outdoor air intake at the air handler is important for providing the correct amount of outdoor airflow to a building. In particular, it has been a common practice for designers to use fixed minimum outdoor air dampers. However, this approach does not necessarily provide good control of outdoor air intake rates, particularly in VAV systems.

In most systems, it is difficult to accurately measure outdoor airflows at the outdoor air dampers during balancing, Cx, or operation. As a result, both overventilation and underventilation can commonly occur. Furthermore, in occupied buildings, overventilation is common since occupancy rates per floor area in most buildings are less than design values. It is estimated that the current amount of energy for ventilating U.S. buildings could be reduced by as much as 30% (first order estimate of savings potential) if the average minimum outdoor rate is reduced to meet the current standards (Fisk et al. 2005).

Accurate measurement of airflows in ducts also requires careful design, proper Cx, and ongoing verification. Under carefully controlled laboratory conditions, commercially available airflow sensors are very accurate. However, in most cases, laboratory conditions and accuracies cannot be replicated in the field; therefore, appropriate correction factors in the programming of the controls may be required.

Continuous monitoring of the outdoor rates at the air handler does not guarantee that the proper amount of ventilation is delivered locally within the building. Poor air mixing both in the ductwork and in the occupied space, especially in larger and more complex air distribution systems, can result in parts of a building receiving less than the design minimum amount of ventilation.

### Measuring Outdoor Airflow

- *Straight Ducts.* Accurate airflow measurements require long, straight duct runs. This presents a challenge to the designer because space and architectural constraints often limit achieving sufficient straight duct lengths.

- *VAV Systems.* VAV systems with single outdoor air intakes need to be designed with modulating dampers and with airflow sensors appropriate for the expected airflow range. In VAV systems with airside economizers, a separate minimum outdoor air intake duct with airflow sensors and a dedicated outdoor air fan with speed control can help ensure accurate control and measurement of the outdoor airflow.

- *Small Packaged Systems.* Small packaged HVAC systems typically do not have continuous measurement of outdoor airflows. This suggests the need for even greater attention to confirmation of the delivery of design airflow rates through balancing, Cx, and periodic recommissioning. For small packaged HVAC systems, straight runs of ductwork in both the supply and return airstreams provide more accurate airflow measurements. Assuming that there is no exhaust (or relief) in the HVAC system, outdoor airflows can then be estimated by subtracting the return airflow rate from the supply airflow rate. Caution needs to be exercised when taking the difference between supply and return airflow measurements in small packaged HVAC systems without sufficient straight ductwork for the supply and return airstreams; such measurements may not meet reasonable accuracy requirements due to cumulative errors in airflow measurement, especially when the outdoor airflow rate is small relative to supply and return airflow rates. If practical, adding ductwork onto the unit's outdoor air intake allows for a traverse of outdoor air.

- *Placement of Sensors.* In general, the best accuracies can be expected when sensors are placed within the manufacturer's guidelines and field-verified for optimum performance. Some research has shown that accuracies of certain measurement technologies may be improved when installed in the following

7.2

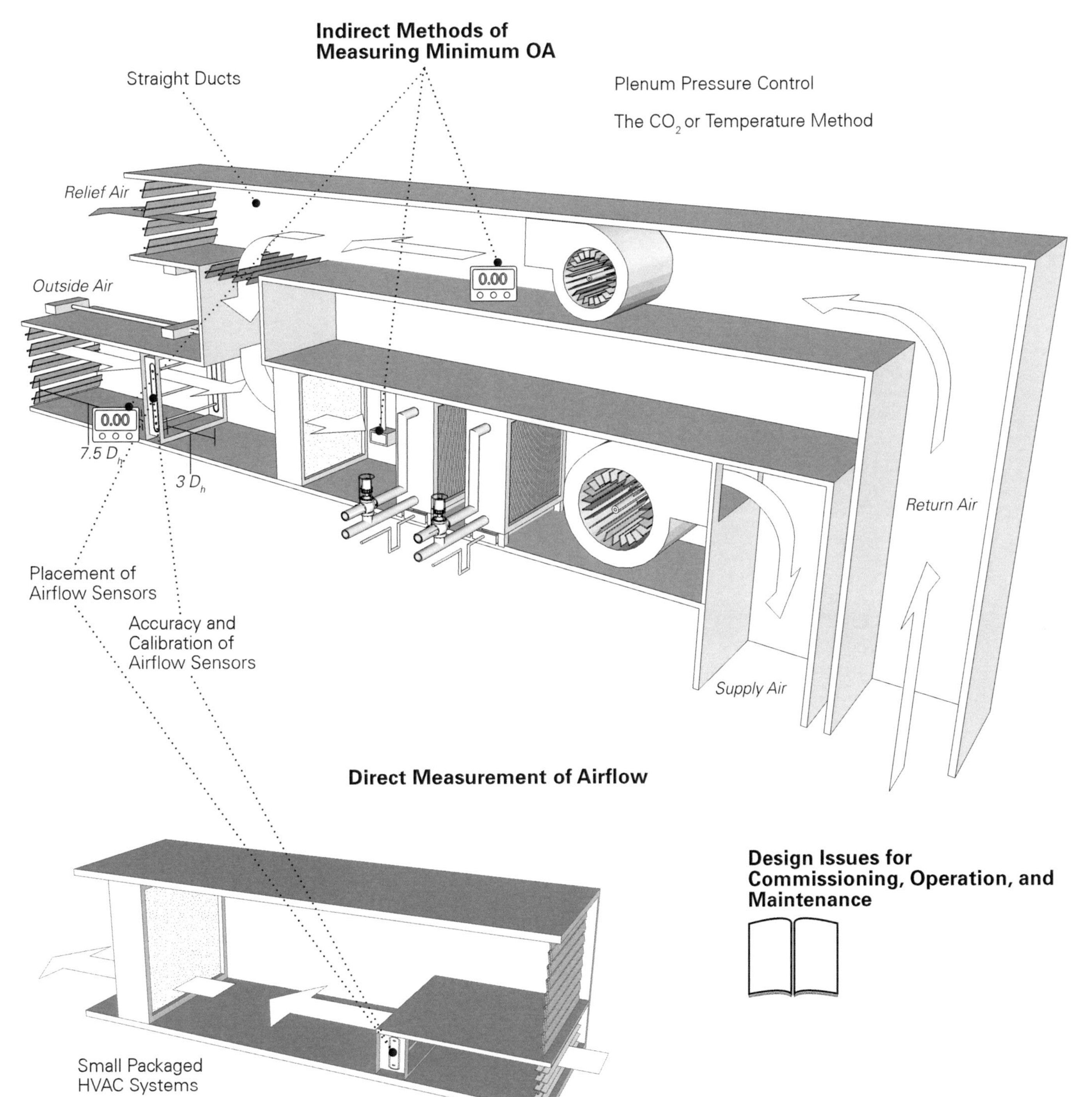
Indirect Methods of
Measuring Minimum OA
Straight Ducts
Plenum Pressure Control
The $CO_2$ or Temperature Method
Relief Air
Outside Air
0.00
0.00
7.5 $D_h$
3 $D_h$
Return Air
Placement of
Airflow Sensors
Accuracy and
Calibration of
Airflow Sensors
Supply Air
Direct Measurement of Airflow
Design Issues for
Commissioning, Operation, and
Maintenance
Small Packaged
HVAC Systems
HVAC Systems with
Economizers

locations: a) between the fixed louver blades where the air speeds are more uniform compared to air speeds downstream of the louvers or b) at the outlet face of the louvers (Fisk et al. 2008). Limited research has shown that in some applications, installation of airflow or pressure sensors downstream of the louvers and upstream of the dampers in combination with an airflow straightening device between the louvers and the airflow or pressure sensors may result in inaccurate airflow measurements (Fisk et al. 2008). Regardless of whether or not airflow sensors are factory or field installed, accuracies of these sensors need to be verified with appropriately calibrated equipment at start-up and during occupancy on regular time intervals.

### Indirect Measurement Methods

Direct measurement methods for measuring outdoor airflow rates are considered to be substantially more accurate than indirect methods. Indirect methods for measuring outdoor airflow rates include plenum pressure control, $CO_2$ concentration balance, $CO_2$ mass balance, supply/return differential calculation, variable-frequency-drive-controlled fan slaving, adiabatic proration formulae, and fixed minimum position intake dampers.

### Design Issues for Commissioning and O&M

The designer needs to make provisions for measurement and verification of the minimum outdoor airflows during the initial Cx as well during the ongoing Cx of a building. Such provisions include easy access to the airflow sensors, hardware and software that can detect sensor (e.g., airflow) and equipment (damper motor) malfunctions, etc. In addition, the design criteria and occupancy assumptions need to be listed in a clear format in the O&M manual so that one can evaluate the continued relevance of design outdoor airflow rates. The building maintenance staff needs to be informed of the need to adjust the minimum amount of outdoor air as space use and occupancy change.

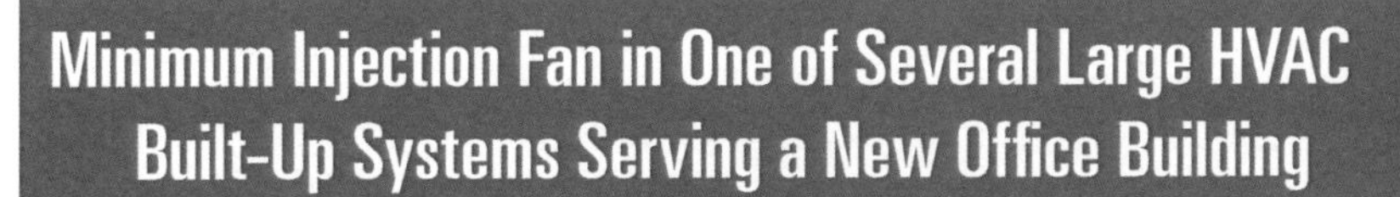

## Minimum Injection Fan in One of Several Large HVAC Built-Up Systems Serving a New Office Building

In order to ensure that the minimum outdoor airflow is provided when the economizer (i.e., 100% outdoor air) is not on, the minimum outdoor air fan in this system is designed to operate when the main outdoor air dampers are closed (Figure 7.2-A). The economizer does not operate when the outdoor air temperature exceeds the return air temperature. A number of identical systems are serving this five-story office building with underfloor air supply in four of the five floors.

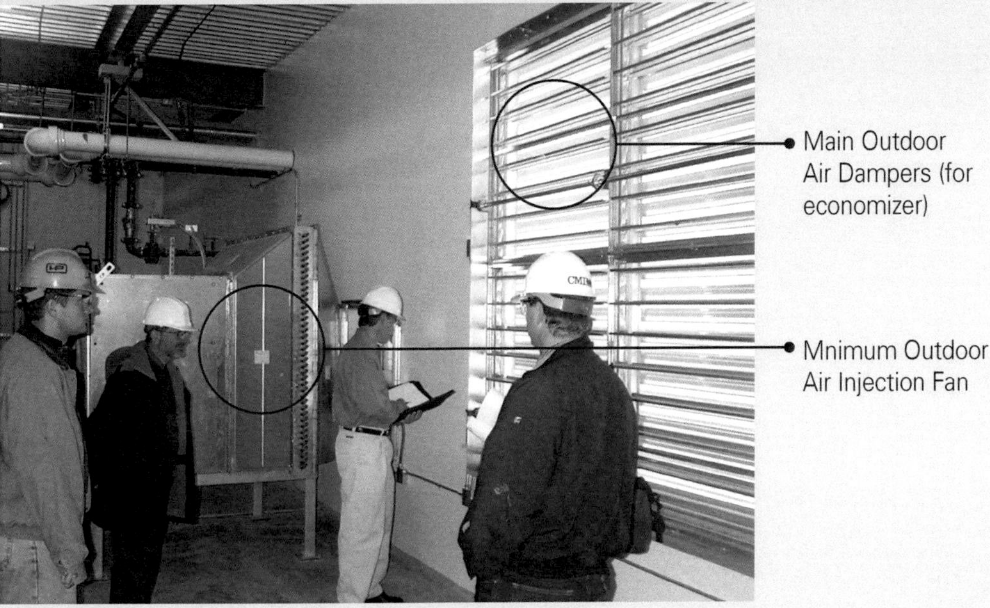


**Figure 7.2-A** Minimum Outdoor Air Fan and Main Outdoor Air Dampers of a Build-Up HVAC System
*Photograph courtesy of Leon Alevantis.*

# Effectively Distribute Ventilation Air to the Breathing Zone



Ventilation only works when the air is delivered to the breathing zone. Different methods of distribution have different efficiencies. For an inefficient system, the quantity of outdoor air at the air handler needs to be increased in order to provide the required minimum quantities in the breathing zone that are required by code and by ASHRAE Standard 62.1.

**Zone Air Distribution Effectiveness**

The airflow rate that needs to be distributed to a zone varies by the effectiveness of the distribution within the room. The ventilation airflow rate provided to the zone needs to be sufficient to provide the required ventilation air to the breathing zone.

The zone outdoor airflow is given by Equation 6.2 in ASHRAE Standard 62.1, as follows:

$$V_{oz} = V_{bz} / E_z$$

where

$V_{oz}$ = quantity of ventilation air delivered to the occupied zone, cfm (L/s)

$V_{bz}$ = quantity of ventilation air delivered to the breathing zone, cfm (L/s)

$E_z$ = zone air distribution effectiveness

Thus, the less efficient an air distribution system is within a zone, the greater will be the required flow of outdoor air to the zone. Choosing air distribution configurations that improve effectiveness, or at least don't decrease it, is therefore an important design decision.

7.3

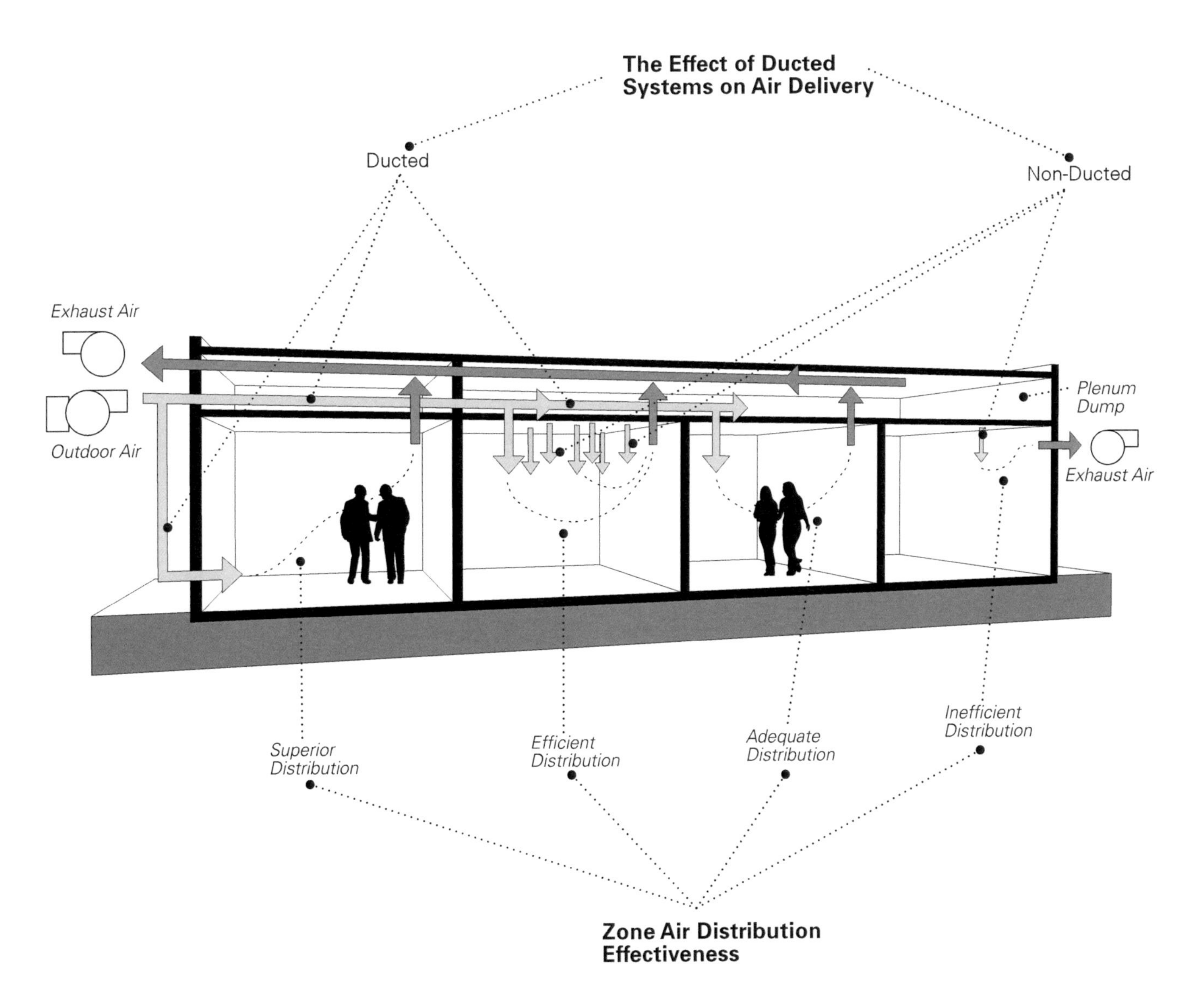
The Effect of Ducted Systems on Air Delivery
Ducted
Non-Ducted
Exhaust Air
Outdoor Air
Plenum Dump
Exhaust Air
Superior Distribution
Efficient Distribution
Adequate Distribution
Inefficient Distribution
Zone Air Distribution Effectiveness

## Zone Air Distribution Effectiveness

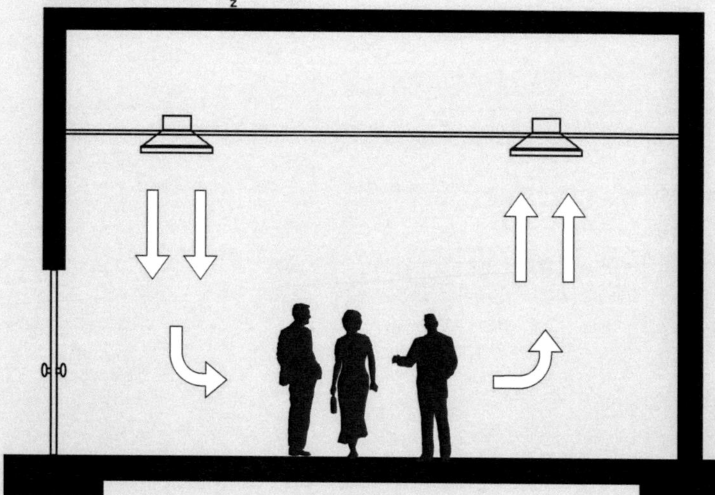


**Figure 7.3-A** Cold Air Falls and is Well Mixed

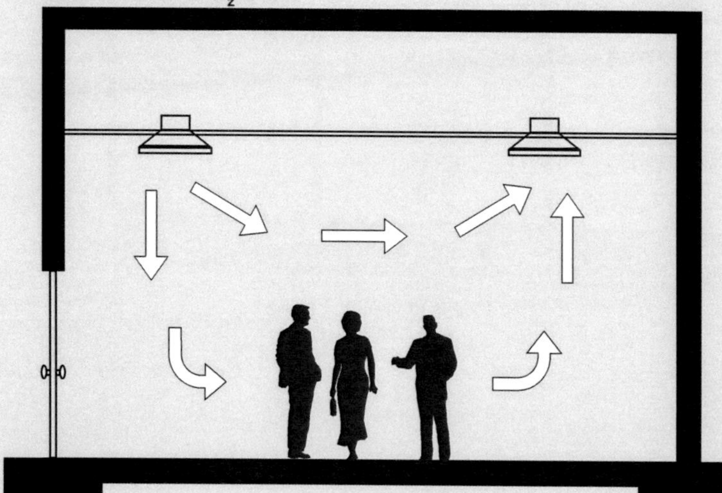


**Figure 7.3-B** Some Hot Air Stays at the Ceiling so Only 80% of the Air Gets to the Breathing Zone

Research in the 1970s revealed that overhead heating with temperatures greater than 15°F (8°C) above ambient resulted in excessive stratification (Int-Hout 2007). Other air distribution configurations can result in varying levels of zone air distribution effectiveness (Figures 7.3-A and 7.3-B). Zone air distribution effectiveness values for varying zone distribution configurations is provided in Table 6.2 of ASHRAE Standard 62.1.

The following are examples of the effect of zone distribution configurations. The effectiveness is categorized here as superior, effective, adequate, and ineffective for presentation clarity.

**Superior Distribution: $E_z = 1.2$**

- Floor supply of cool air and ceiling return, provided when low-velocity displacement ventilation achieves unidirectional flow and thermal stratification

**Effective Distribution: $E_z = 1.0$**

- Ceiling supply of cool air
- Ceiling supply of warm air and floor return
- Ceiling supply of warm air less than 15°F (8°C) above space temperature and ceiling return provided that the 150 fpm (0.8 m/s) supply air jet reaches to within 4.5 ft (1.4 m) of floor level.
- Floor supply of cool air and ceiling return provided that the 150 fpm (0.8 m/s) supply jet reaches 4.5 ft (1.4 m) or more above the floor
- Floor supply of warm air and floor return

**Adequate Distribution: $E_z = 0.8$.**

- Ceiling supply of warm air 15°F (8°C) or more above space temperature and ceiling return
- Makeup supply drawn in on the opposite side of the room from the exhaust and/or return

**Ineffective Distribution: $E_z < 0.8$**

- Floor supply of warm air and ceiling return
- Makeup supply drawn in near the exhaust and/or return location

STRATEGY OBJECTIVE 7.3

# Effectively Distribute Ventilation Air to Multiple Spaces

Ventilation only works when the air is delivered to the breathing zone. Different methods of distribution have different efficiencies. For an inefficient system, the outdoor airflow rate at the air handler must be increased in order to provide the required minimum outdoor airflow rate to the breathing zone. For multiple-zone recirculating systems, the system will have an efficiency ($E_v$) that needs to be calculated to determine the outdoor airflow rate required at the air handler. This efficiency is for the system and needs to be used in addition to the corrections for effectiveness of distributing air within the zone ($E_z$). The values for system efficiency can range from 1.0 to 0.3 or lower, with higher values being more efficient (better).



## System Ventilation Efficiency

**Figure 7.4-A** Building where Multiple-Space Equations Were Tested
*Photograph courtesy of Yuill et al. (2007).*

### System Ventilation Efficiency

For a single-zone system or a DOAS, the airflow rate at the outdoor air intake ($V_{ot}$) is equal to the air required for the zone(s). The numbered equations are taken from ASHRAE Standard 62.1-2007, including Appendix A of that standard.

$$V_{ot} = V_{oz} \quad (6\text{-}3)$$

$$V_{ot} = \sum_{all\ zones} V_{oz} \quad (6\text{-}4)$$

For systems that supply multiple zones, the calculations are more complex. The airflow rate required at the outdoor air intake is driven by the critical zone. However, the percent outdoor air required at the system intake ($X_s$) can be less than the percent outdoor air required at the critical zone ($Z_d$ critical). The system ventilation efficiency is given by the following equation:

Single Supply Systems $\quad E_{vz} = 1 + X_s - Z_d \quad$ (A-1)

The multiple-space equations were tested in the building illustrated in Figure 7.4-A, and a research project confirmed that the system intake air fraction does not have to be as large as the fraction of outdoor air at the critical zone in order to deliver the proper outdoor airflow rate ($V_{oz}$) to the zone (Yuill et al. 2007).

# Strategy 7.4

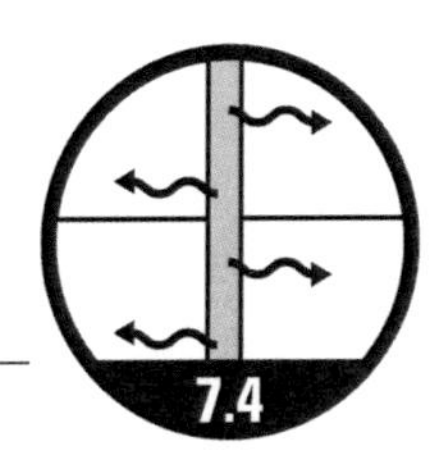


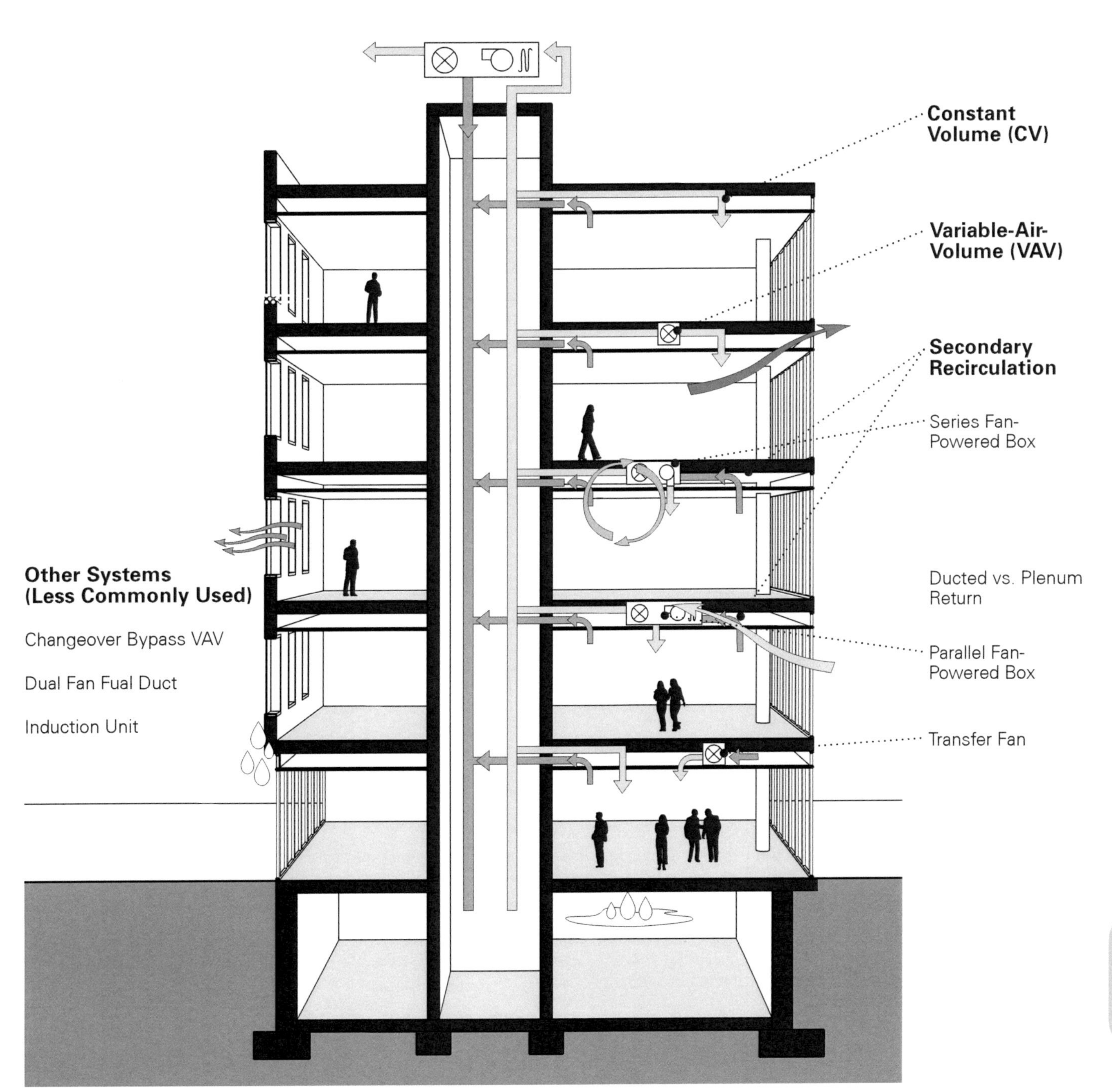


STRATEGY OBJECTIVE 7.4

# Provide Particle Filtration and Gas-Phase Air Cleaning Consistent with Project IAQ Objectives

Filtration and gas-phase air cleaning (FAC) strategies are important because they can provide a more effective means of removing contaminants than source control or ventilation in some cases. In this way, filtration and air cleaning can help reduce occupant exposure to a variety of contaminants and thus improve occupant health, comfort, and productivity.



The technologies of enhanced particulate FAC are well developed, with a history of over half a century of application in industrial process control, cleanrooms, health care and pharmaceuticals, and special usage buildings such as museums and laboratories. However, filtration systems have not been widely applied in general commercial buildings. This Strategy provides a broad understanding and appreciation of the potential of FAC as a tool for attaining enhanced levels of IAQ. Both particulate and gas-phase control are discussed together in this Strategy because much of the design consideration and mechanical application technology is concurrent and co-mingled.

A full range of particulate FAC equipment is available from a wide selection of manufacturers. The filter cartridges vary widely in frame methodology and seal mechanisms, filter/sorption media, cartridge depth and size, loading characteristics and capacity, surface area configurations, airflow and pressure drop, and cost. The efficiency or removal performance is a critical component of the selection of FAC equipment. However, other factors also influence the overall performance and total value of the FAC system and, thus, should be included in the selection and specification process.

The application of FAC is a balance of:

- the control and extraction requirements,
- the physical/mechanical limitations of the air-handling system,
- the characteristics of the CoC, and
- the features of the FAC equipment.

Generally, the space requirements, energy, and O&M costs increase with higher efficiency requirements. Because of cost and space constraints, the final selection of equipment should go beyond removal efficiency to include capacity and life cycle, maintenance requirements, and lifelong pressure drop factors. Because of these factors, life-cycle cost, including energy requirements, should drive the selection of FAC rather than just first cost. The supportive discussion in the Part II detailed guidance to this Strategy in the electronic version of this Guide provides rationale for the selection of specific extraction efficiencies and specific FAC equipment along with specific recommendations.

Early consideration should be devoted to FAC in a number of overall building design aspects. This early consideration is necessary because the function of FAC equipment is similar to ventilation systems except the FAC system employs extraction rather than dilution to lessen the concentration of unwanted contaminants in the conditioned space. Thus, it can be applied in conjunction with ventilation to aid and improve the quality of the ventilation air or as an alternative or adjunct to ventilation air for economic reasons. Because of these factors, FAC has interactivity features that are similar to ventilation, including the evaluation of the outdoor air quality, an understanding of internal sources, knowledge of occupant density and activity, concern for air capture and recirculation efficacy, selection of HVAC equipment, and concern for energy management issues.

# Strategy 7.5

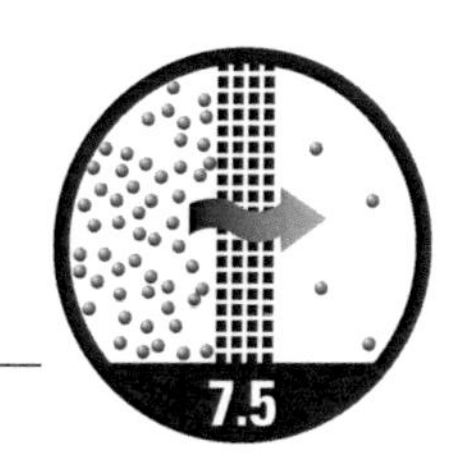


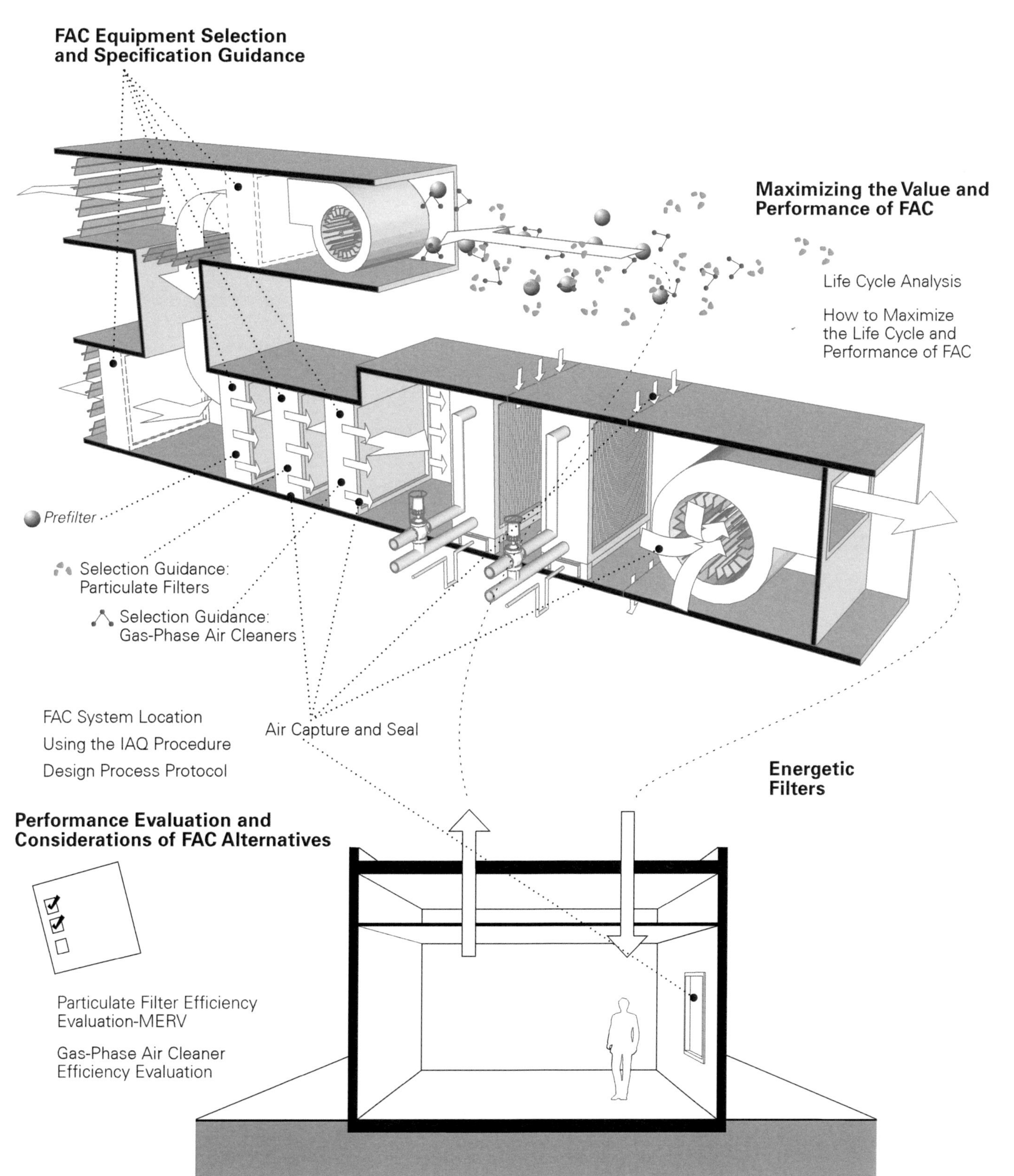


STRATEGY OBJECTIVE 7.5

The following are typical interrelated areas that trigger early consideration of FAC by the design team:

- *When outdoor ventilation air is unreliable* for delivering dependable and effective dilution of indoor contaminants because of preexisting pollutant content, such as ozone or ultrafine particulate matter, or when other odor or pollutant generating sources are nearby.
- *When outdoor ventilation air is either consistently or seasonally burdened with high heat and latent loads* demanding additional and excessive HVAC equipment capacity and operating energy consumption.
- *When source control tactics, such as material selection and localized exhaust, are insufficient* for adequately lowering concentrations of CoC.
- *When occupant function or activity generates elevated levels of CoC* that cannot be adequately controlled by routine levels of dilution ventilation.
- *When internal occupancy patterns include high density and/or wide diversity of space usage* requiring high peak ventilation levels with related capacity and energy load.
- *When the owner or tenant needs or expectations require higher levels of contaminant control* to provide enhanced levels of indoor environmental acceptability and conditions.
- *When occupants, occupant activities, or contents of the building require enhanced levels of protection* from the deleterious effects of airborne contaminants, whether the CoC are microbiological, chemical, or particulate in nature.
- *When exhaust systems contain CoC that have the potential for re-entrainment* or contribution of potential risk to the ambient air.

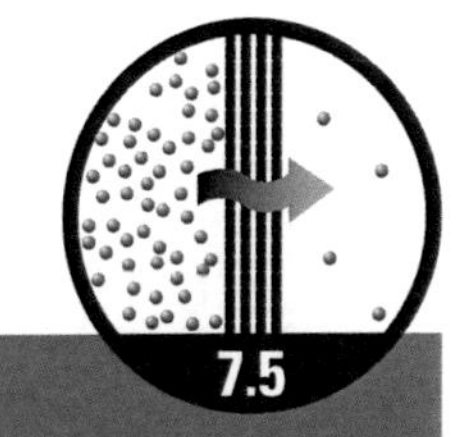


## Improved IAQ and Protected Objects in Specialty City Museum

The low-rise building shown in Figure 7.5-A serves the archival offices and laboratories of a museum complex located in a major Southeastern city. The building is less than ten years old and is used for preservation and restoration of historically significant photos, documents, and objects d'art that are mostly cellulosic and/or organic-based in content. Thus, there are a number of odorous, noxious, and potentially hazardous chemicals used in the various preservation processes employed by the museum professionals. These normally require a high level of ventilation and dilution to maintain acceptable IAQ. However, the same historical documents are highly susceptible to the gaseous contaminants of ambient outdoor urban air. Thus, the facility HVAC system operates with reduced outdoor air, and both the outdoor air and return airstreams are treated with filtration and gaseous air cleaning to remove the chemical contaminants from both the outdoor air and the return air. This results in acceptable indoor environmental conditions for both the occupants and the valuable stored materials while reducing both the risk and the cost of introducing excessive quantities of outdoor air.

**Figure 7.5-A** Museum Archival Offices
*Photograph courtesy of H.E. Burroughs.*

The special FAC equipment consists of MERV 6 prefilters and 12 in. deep medium efficiency matrix media imbedded gas-phase filter cartridges and is located in the mixed air plenum to treat both return and outdoor air sources. On-site performance testing conducted in 2007 revealed that the system efficiency for particulate reduction was 94.6% at 0.5 µm size particles and the TVOC levels were maintained below 70 µg/m$^3$.

# Provide Comfort Conditions that Enhance Occupant Satisfaction

Thermal conditions indoors, combined with occupant activity and clothing, determine occupant thermal comfort, which in turn impacts occupant productivity and perceptions of air quality. Dry-bulb temperature is only one physical parameter out of many that interact in a complex manner to produce occupant satisfaction.

Thermal conditions affect chemical and biological contaminant levels and/or the intensity of occupants' reactions to these contaminants, but our knowledge of these effects and their mechanisms is very limited. Despite this limited knowledge, achieving high performance in thermal comfort is likely to result in lower contaminant levels and better occupant perceptions of IAQ.



Tools exist for calculating the proportion of people likely to be satisfied by the combination of comfort factors that include dry-bulb temperature, humidity, air velocity, and radiant temperature. The most commonly known tool in the U.S. is *ASHRAE Thermal Comfort Tool* (ASHRAE 1997), a computer program that is part of *ANSI/ASHRAE Standard 55, Thermal Environmental Conditions for Human Occupancy* (ASHRAE 2004).[1] In order to use this tool, the amount of clothing worn by occupants and their levels of metabolic (or physical) activity must be provided, in units of clo (clothing level) and met (metabolic rate), respectively.

In traditional designs, the HVAC designer's role in achieving comfort conditions often begins and ends at the selection of an indoor design condition and the sizing of the HVAC system to provide these conditions at peak load. The selection of a design dry-bulb condition involves both comfort and cost or energy considerations and can dictate critical design features of the system. For example, some designers may pick a relatively high design cooling condition such as 78°F (26°C) in order to conserve energy, while others may select one such as 72°F (22°C) in order to maximize the number of satisfied occupants. Selecting systems and controls that perform efficiently at part load can mitigate the energy downside of the latter.

Each person having control over his or her own environment, referred to as *personalized ventilation and conditioning* (as provided in many automobiles and airplanes, for example) is the ideal situation but is not easily attained in buildings. It is wise, therefore, to select zones carefully and consider using as many as is needed to create sufficient homogeneity within each zone to improve the ability to satisfy comfort needs of occupants in the zones. Rooms and areas having loads that vary over time in patterns that are significantly different from areas that surround them benefit from having their own conditioning control loops and thermostats.

It is expected that individual occupants in the same temperature control zone will have different thermal comfort needs. They can be encouraged, therefore, to adjust their clothing to fit their own needs. However, if the occupants have an adjustable thermostat in the space, then so-called "thermostat wars" may occur, where occupants frequently readjust the thermostat that others have set. This situation can reduce both the efficiency and effectiveness of the comfort system. The solution in some cases may be either giving control to a neutral party, such as the building operator or office manager, or using thermostats for which the temperature adjustment range is limited.

The control of humidity at part load is a comfort goal that needs to be considered in the design of systems and their control sequences. Controlling moisture is also important to limit condensation and mold, as discussed in Strategy 2.4 – Control Indoor Humidity.

Air diffusion devices need be selected so that the required air velocity conditions in occupied zones are maintained at low airflow, as would occur in a VAV system. It is also important to choose thermostat locations that best represent the conditions that occupants will experience and that are not confounded by solar radiation or other heat sources.

[1] The program code is published in Normative Appendix D of ASHRAE Standard 55-2004 in C++/BASIC. A user-friendly version on CD is available for purchase from ASHRAE.

# Strategy 7.6

7.6

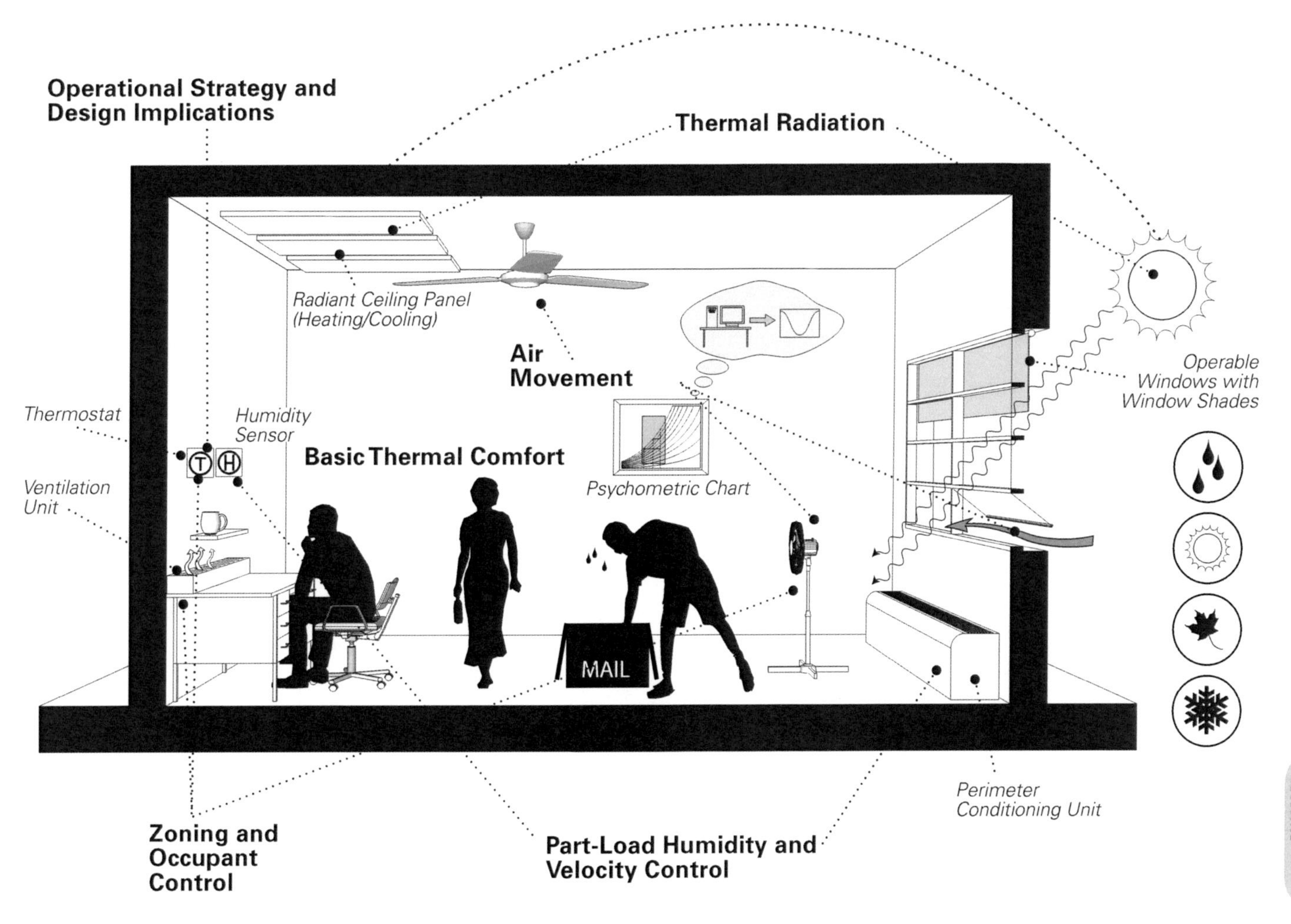
Operational Strategy and Design Implications
Thermal Radiation
Radiant Ceiling Panel (Heating/Cooling)
Air Movement
Operable Windows with Window Shades
Thermostat
Humidity Sensor
Basic Thermal Comfort
Ventilation Unit
Psychometric Chart
MAIL
Perimeter Conditioning Unit
Zoning and Occupant Control
Part-Load Humidity and Velocity Control

Taking thermal radiation into account in system capacity can mitigate the negative effects of strong radiant sources while the comfort benefits—for instance, radiant heating—can be realized.

Drafts can occur under windows and other concentrated locations of heat loss. Overhead forced air for heating can be low in cost, but its effectiveness in avoiding drafts and comfort diminishes with increased height and temperature.

## Layout of a Corridor with Large Amounts of Glazing

In order to achieve visual symmetry on the outside of the building, the corridor shown in Figure 7.6-A has 7 ft (2.1 m) high glazing. Because the building faces southeast, it requires massive amounts of air for cooling, which in turn increases both first costs and energy costs.

Discussions are under way with the design team to make one or several of the following improvements:

- reduce the amount of glazing,
- use glass with a heavy shading coefficient,
- install interior shading,
- use natural conditioning by making the windows operable and isolating the corridor from mechanically conditioned spaces, and/or
- relax the comfort criteria in the corridor since people pass through it quickly.

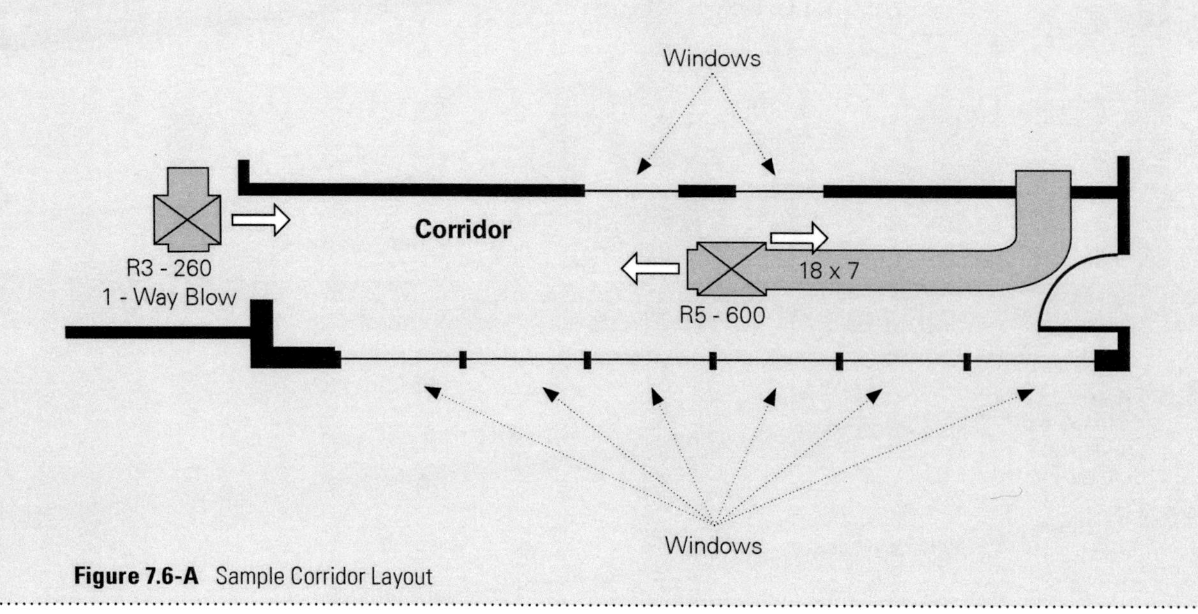


**Figure 7.6-A** Sample Corridor Layout

7.6

STRATEGY OBJECTIVE 7.6

# Apply More Advanced Ventilation Approaches

Conditioning and transporting ventilation air accounts for a significant fraction of building energy use. The Strategies presented in this Objective can help reduce the energy required to deliver good IAQ.

- Strategy 8.1 – Use Dedicated Outdoor Air Systems Where Appropriate covers systems that condition 100% outdoor air and deliver it directly to occupied spaces or to other heating/cooling units that serve those spaces. DOASs can make it easier to verify that the required amount of outdoor air is delivered and can reduce the total outdoor air required relative to other systems. DOASs can easily be combined with energy recovery or DCV to further reduce energy use.
- Energy recovery ventilation reduces energy use by transferring energy from the exhaust airstream to the outdoor airstream. Strategy 8.2 – Use Energy Recovery Ventilation Where Appropriate explains when energy recovery ventilation is required by energy standards and when it can have favorable economics even though not required as well as how it can improve humidity control and reduce the risk of mold growth.
- DCV varies ventilation airflow based on measures of the number of occupants present. It can be particularly cost-effective for spaces with intermittent or highly variable occupancy. Strategy 8.3 – Use Demand-Controlled Ventilation Where Appropriate describes DCV design concepts and considerations.
- Natural ventilation can be a low-energy strategy that provides a pleasant environment in mild climates with good outdoor air quality. Mixed-mode ventilation can provide similar benefits in additional climates through the limited use of mechanical equipment. Meeting ventilation requirements with natural or mixed-mode ventilation is a new challenge that requires careful design, as described in Strategy 8.4 – Use Natural or Mixed-Mode Ventilation Where Appropriate.
- Strategy 8.5 – Use the ASHRAE Standard 62.1 IAQ Procedure Where Appropriate describes an alternative to the Ventilation Rate Procedure that can be used to comply with ASHRAE Standard 62.1 (ASHRAE 2007a), using lower outdoor airflow rates or to provide enhanced IAQ with the same rates. The IAQ Procedure can be cost-effective in applications requiring large volumes of ventilation air in climates where the cost to condition outdoor air is high.

Before considering these approaches, it is important to understand the basic issues described in Objective 7 – Reduce Contaminant Concentrations through Ventilation, Filtration, and Air Cleaning.

# Objective 8

**8.1** Use Dedicated Outdoor Air Systems Where Appropriate

**8.2** Use Energy Recovery Ventilation Where Appropriate

**8.3** Use Demand-Controlled Ventilation Where Appropriate

**8.4** Use Natural or Mixed-Mode Ventilation Where Appropriate

**8.5** Use the ASHRAE Standard 62.1 IAQ Procedure Where Appropriate

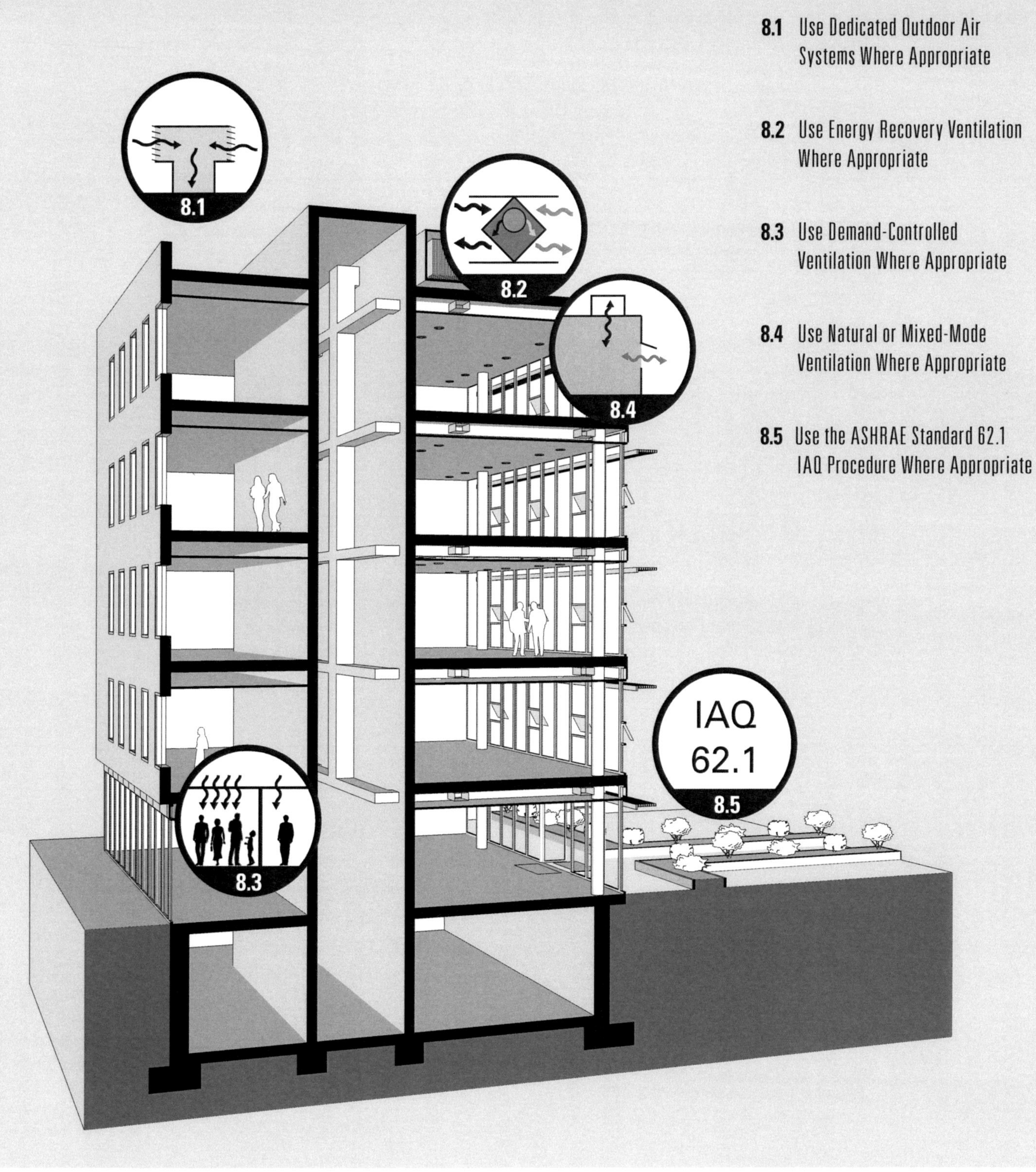

# Use Dedicated Outdoor Air Systems Where Appropriate

All DOASs are 100% outdoor air systems. The DOAS approach makes calculating the required outdoor ventilation airflow more straightforward than for multiple-space systems. Having the ventilation system decoupled from the heating and air-conditioning system can provide many advantages for HVAC system design. A disadvantage may be that there is an additional item of equipment, the DOAS unit itself.

DOASs must address latent loads, the largest being the latent load from the outdoor air in some cases. The DOAS may also be designed to remove the latent load from both the outdoor air and the building (total latent load), in which case there are multiple advantages.

If the exhaust airstream is located close to the ventilation airstream, both sensible and latent energy can be recovered in the DOAS. This feature makes DOASs much more energy efficient. It is not necessary that the exhaust and supply airflows be exactly the same rate, but if they differ the difference must be accounted for in the equipment sizing calculations.v



**DOAS Component Combinations**

A DOAS is made up of a site-appropriate selection of components and can be built-up or manufactured. In most areas of the country, cooling coils (CCs) are required to cool and dehumidify the air. In some areas, heating coils may be required.

Integration of energy recovery technology can reduce the load on the heating and cooling coils. Energy recovery components can be either total (enthalpy) energy recovery or sensible energy recovery. See Strategy 8.2 – Use Energy Recovery Ventilation Where Appropriate.

Because of the latent load of outdoor air, in many areas use of an active desiccant wheel (AdesW) or a passive dehumidification component (PDHC) may be cost justified. These devices assist in managing humidity within the building. The rationale for humidity control is presented in Strategy 2.4 – Control Indoor Humidity.

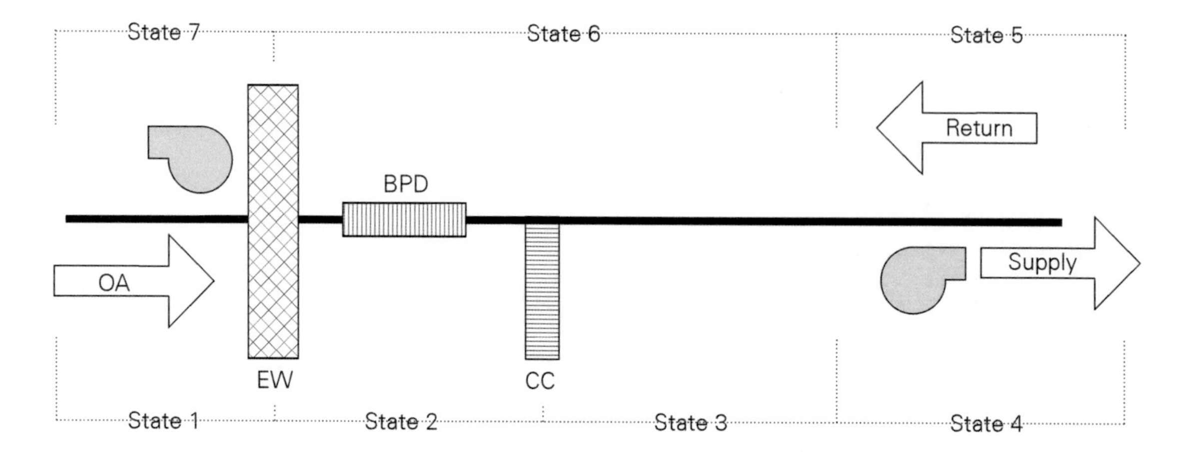


**Figure 8.1-A** Example of DOAS with enthalpy wheel (EW) and CC

# Strategy 8.1

8.1

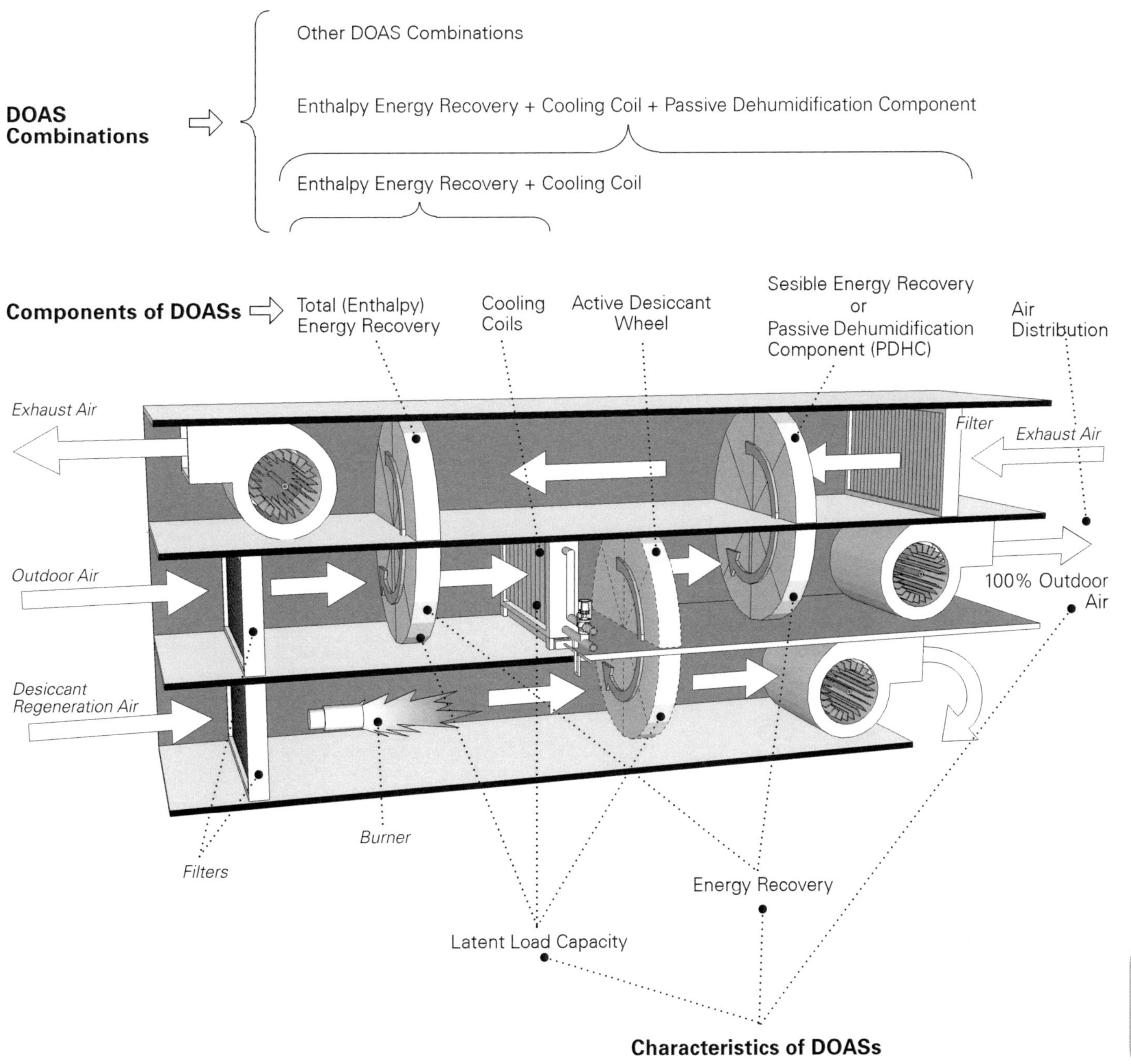
DOAS Combinations
Other DOAS Combinations
Enthalpy Energy Recovery + Cooling Coil + Passive Dehumidification Component
Enthalpy Energy Recovery + Cooling Coil
Components of DOASs
Total (Enthalpy) Energy Recovery
Cooling Coils
Active Desiccant Wheel
Sesible Energy Recovery or Passive Dehumidification Component (PDHC)
Air Distribution
Exhaust Air
Filter
Exhaust Air
Outdoor Air
100% Outdoor Air
Desiccant Regeneration Air
Burner
Filters
Energy Recovery
Latent Load Capacity
Characteristics of DOASs

After the air is conditioned by the DOAS, it must still to be delivered to the space. The design must specify how the air is to be delivered and how it is to interact with other heating and cooling equipment located in the spaces.

- *Enthalpy Energy Recovery + Cooling Coil.* A straightforward and efficient DOAS can be constructed with a CC and an enthalpy air-to-air energy recovery device when the exhaust airstream is available for energy recovery (see Figure 8.1-A).
- *Enthalpy Energy Recovery + Cooling Coil + Passive Dehumidification Component.* Another efficient DOAS can be constructed using a CC, passive dehumidification, and enthalpy energy recovery. This system also requires that an exhaust airstream be available (see Figure 8.1-B).
- *Other DOAS Combinations.* There are many other combinations of components that are appropriate for differing applications. What is site appropriate depends on local climate, availability of waste heat to regenerate desiccants, and many other factors.

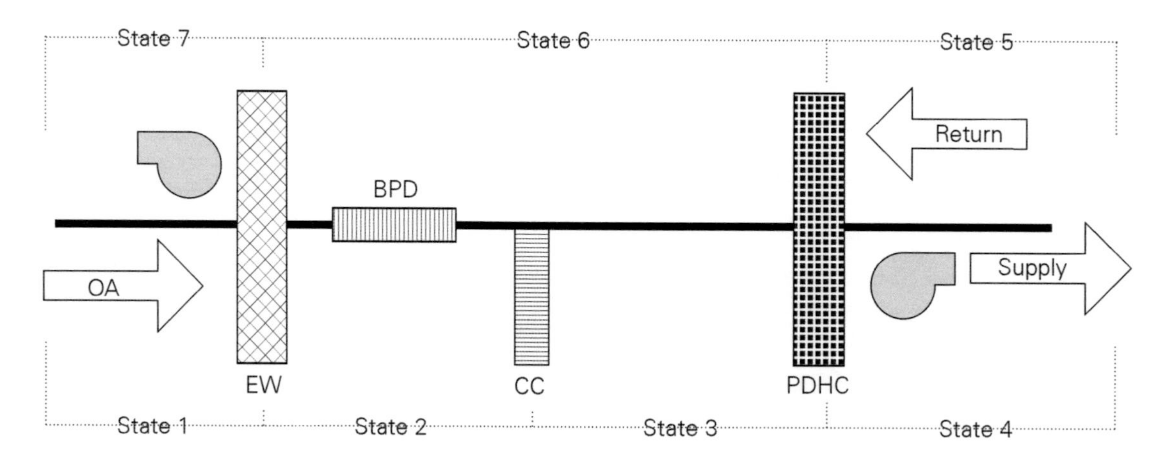


**Figure 8.1-B** Example of DOAS with EW, CC, and PDHC

8.1

## Junior High School in Pennsylvania with DOAS

**Figure 8.1-C** Junior High School in Pennsylvania (Roof-Mounted DX DOAS Visible)
*Photograph copyright McClure Company.*

The Halifax Elementary School (Figure 8.1-C) is a 53,450 ft$^2$ (4,946 m$^3$) building in Central Pennsylvania. The original facility contained a traditional two-pipe unit ventilator heating and air-conditioning system. Due to storm water control problems and the inability of the HVAC systems to properly dehumidify the facility, the school experienced excessively high relative humidity conditions.

In 2006 the building HVAC and electrical systems were upgraded. Packaged gas-electric rooftop units with enthalpy heat recovery wheels were installed to provide dedicated outdoor ventilation air to the classrooms. The unit ventilators were removed and replaced with two-pipe fan-coil units, which are decoupled from the outdoor ventilation air. The unit ventilator outdoor air intakes were tightly covered with insulated panels. The fan-coil units satisfy individual room heating and sensible cooling requirements. Decoupling the DX DOAS from the two-pipe fan-coil system greatly simplified the semi-annual changeover between hot and chilled water distribution.

The HVAC upgrades were implemented at a cost of \$15.50/ft$^2$ (\$164/m$^2$). Facility energy use was reduced from 105 kBtu/ft$^2$ (1190 MJ/m$^2$) before the project to 97 kBtu/ft$^2$ (1100 MJ/m$^2$) after the project was completed. Sixty-five percent of these energy savings are associated with the HVAC portion of the project, while the remainder is due to lighting system upgrades. The facility no longer experiences excessive relative humidity.

# Use Energy Recovery Ventilation Where Appropriate

Energy recovery ventilation is required for certain applications by energy standards such as *ANSI/ASHRAE/IESNA Standard 90.1, Energy Standard for Buildings Except Low-Rise Residential Buildings* (ASHRAE 2007e). In other cases, energy recovery systems provide such sufficient payback in overall system sizing and reduced operating costs over the life of the system that they are installed voluntarily. In the case of total energy recovery ventilators (ERVs), improved humidity control is an additional benefit, critical for controlling condensation and mold growth and for thermal comfort. In some cases an ERV is the critical component that allows the HVAC system to achieve moisture management objectives.



In general, there are two types of energy recovery ventilation devices: total ERVs that transfer heat and moisture between incoming and exhaust air and heat recovery ventilators (HRVs) that do not transfer moisture. Along with generic IAQ concerns about keeping mechanical air delivery equipment clean, proper application of equipment needs to address correct selection, sizing, application, Cx, and maintenance. Because this equipment needs to perform correctly under all outdoor weather conditions, it is essential that the system containing the energy recovery devices be properly commissioned for the specific application.

Types of energy recovery systems include an energy recovery wheel and fixed plate with latent transfer, fixed plate, heat pipe, and runaround loop systems.

General design considerations include those associated with filtration, controls, sizing, condensation, and sensible heat ratio.

**Warning:** Air-to-air energy recovery equipment needs to have airstreams filtered and/or cleaned for efficient, reliable operation. Like any other equipment, the system needs to be properly maintained.

## Using Energy Recovery System to Control Humidity

In humid climates, there is a challenge in properly ventilating schools and also controlling indoor humidity. ASHRAE Standard 62.1-2007 requires that relative humidity in a space be limited to less than 65% during cooling conditions. Using packaged equipment for air conditioning and ventilation often cannot provide humidity control under part-load conditions. This is because a typical packaged equipment sensible heat ratio is 0.67 when the part-load sensible heat ratio in a classroom is closer to 0.40 (Fischer and Bayer 2003). Using an energy recovery system with latent transfer capability can provide improved humidity control while saving energy. A recent study of an energy recovery system documented reductions in space humidity with corresponding energy savings at schools in the Atlanta area (Fischer and Bayer 2003).

Figure 8.2-A shows sensible and latent loads for a typical classroom in Atlanta, GA.

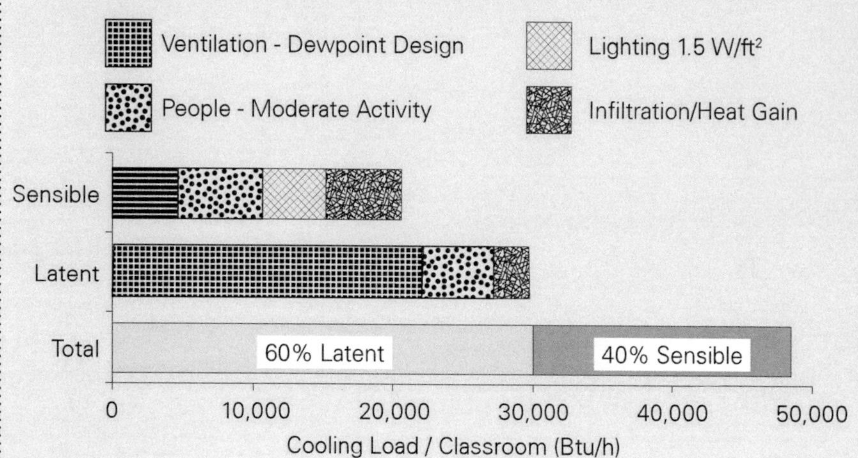


**Figure 8.2-A** Sensible and Latent Loads for a Typical Classroom in Atlanta

# Strategy 8.2

8.2

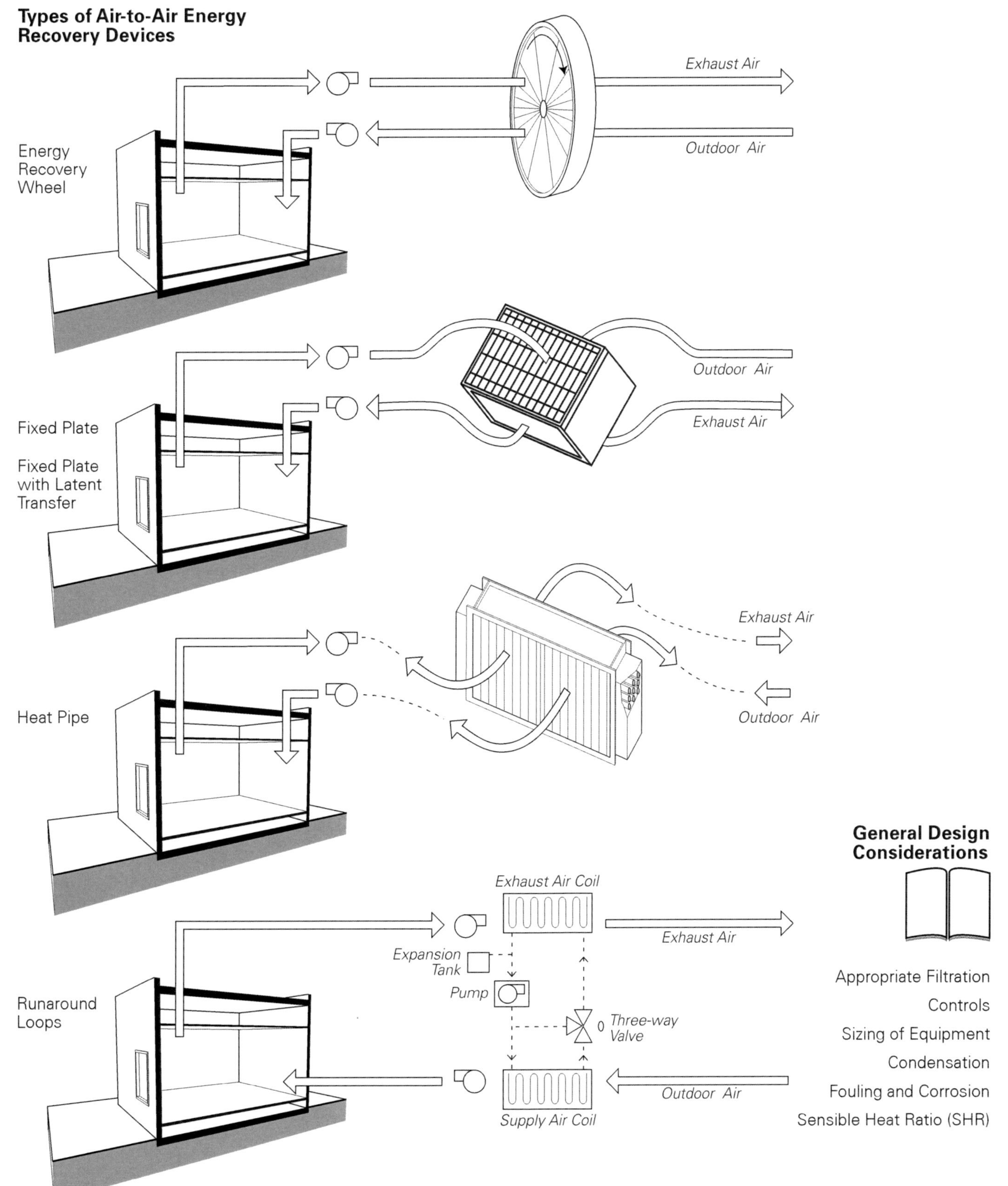


**General Design Considerations**

Appropriate Filtration

Controls

Sizing of Equipment

Condensation

Fouling and Corrosion

Sensible Heat Ratio (SHR)

# Use Demand-Controlled Ventilation Where Appropriate

Demand-controlled ventilation (DCV) is a control strategy that varies the amount of ventilation by resetting the outdoor air intake flow setpoints to an occupied space based on the changing number of occupants. The goal is to avoid underventilation (increasing the potential for poor IAQ) as well as overventilation (wasted energy). It has been estimated that in U.S. commercial buildings, DCV has the potential to reduce heating and cooling loads by as much as 20% or from \$0.05/ft$^2$ (\$0.54/m$^2$) to more than \$1/ft$^2$ (\$11/m$^2$) annually. However, actual savings can vary widely depending on climate, variability in population density and occupancy schedule, type of building, whether or not the HVAC system has an economizer, and other factors.



The simplest approach to DCV is control of the outdoor air rate in an on-off manner based on signals from a room occupancy sensor, time clock, or light switch. A more sophisticated approach uses a signal that is proportional to the number of persons in a space to automatically modulate the amount of outdoor air.

**Appropriate Application of DCV**

DCV is most appropriate in densely occupied spaces with intermittent or variable population. For these spaces, DCV offers the potential for both energy savings and improved IAQ. The benefit of DCV increases with the level of density, transiency, and cost of energy.

Occupancy categories most appropriate for DCV include theaters, auditoriums/public assembly spaces, gyms, some classrooms, restaurants, office conference rooms, etc. Densely occupied spaces with people-related pollutants other than normal bioeffluents, such as waiting areas of health-care facilities, are less appropriate for DCV despite their intermittent or variable population. Densely and continuously occupied office spaces, such as call centers, are less likely to see the benefits of DCV given the lack of variability in occupancy.

Although the energy-conserving benefits of DCV may be small in general office buildings, making DCV a not-very-cost-effective energy-saving strategy in such applications, certain aspects of DCV controls may be beneficial to such a building in ensuring that the design ventilation rates are supplied under all operating conditions. For example, continuous measurement of outdoor airflow rates and indoor $CO_2$ levels can help building personnel find ventilation system faults or make adjustments to the HVAC system setpoints, thus avoiding overventilation or underventilation relative to the design or code requirements.

**DCV in Multiple-Zone Systems**

Application of DCV in single-zone systems is fairly straightforward. However, neither ASHRAE Standard 62.1-2007 nor its associated user's manual address the design and operation of DCV for systems that serve multiple spaces. There is currently no published guidance for DCV in multiple-zone systems.

**$CO_2$-Based DCV**

Measurement and control of indoor $CO_2$ concentrations has been the most popular DCV method because $CO_2$ sensors and associated controllers are relatively inexpensive and, in controlled environments, have been shown to correlate well with people-related contaminant levels (Persily 1997). A number of packaged HVAC equipment manufacturers now offer $CO_2$ sensors and controllers as an option for their equipment. This method is based on the fact that the rate of $CO_2$ generation indoors by occupants is proportional to the number of occupants and their activity levels. Other indoor sources of $CO_2$ and removal mechanisms may exist in some buildings, and in some cases they may be significant enough to compromise the viability of a $CO_2$-based DCV.

*Design Considerations for $CO_2$-Based DCV.* $CO_2$-based DCV is required by some building codes. However, despite its relatively low cost and short payback, the $CO_2$-based DCV market has grown slowly since 1990 and has not necessarily reached its peak potential. This is partially due to the limited data on the long-term performance of

# Strategy 8.3

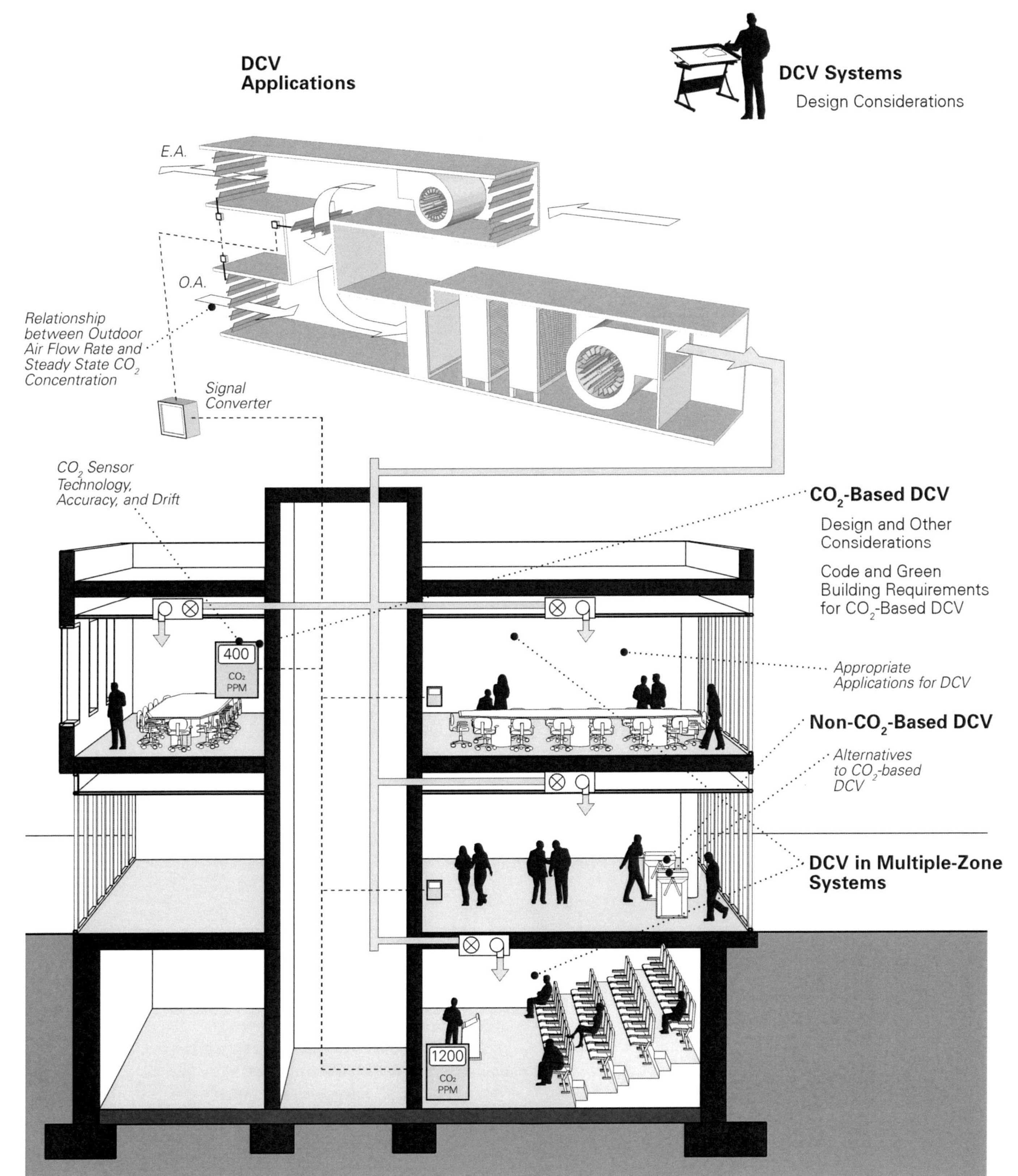
8.3
DCV Applications
DCV Systems
Design Considerations
E.A.
O.A.
Relationship between Outdoor Air Flow Rate and Steady State $CO_2$ Concentration
Signal Converter
$CO_2$ Sensor Technology, Accuracy, and Drift
$CO_2$-Based DCV
Design and Other Considerations
Code and Green Building Requirements for $CO_2$-Based DCV
400
$CO_2$ PPM
Appropriate Applications for DCV
Non-$CO_2$-Based DCV
Alternatives to $CO_2$-based DCV
DCV in Multiple-Zone Systems
1200
$CO_2$ PPM

these sensors. Limited studies have indicated that there are numerous issues that need to be addressed by further research. Some of the reported issues with the $CO_2$ sensors relate to the accuracy of the sensors while others relate to maintenance/calibration and to the sensor lag times (Shrestha and Maxwell 2009a, 2009b, 2010; Emmerich and Persily 2001; Fisk et al. 2007). Also, the $CO_2$ generation rates measured and reported for sedentary adults (1.2 met units) need to be adjusted for other situations, such as children in classrooms.

The following considerations should be made during the design of a $CO_2$-based DCV system:

- In HVAC systems with open plenum returns, $CO_2$ sensors should be located in the room so that the average concentrations at breathing level can be obtained. A sufficient number of sensors should be placed within a space in order to increase the certainty of the sensed average space $CO_2$ concentration. Sensors placed in return air plenums will not necessarily yield a reliable value representative of the average breathing concentration for the space.
- In HVAC systems with ducted returns, $CO_2$ sensors may be placed in the return air duct from a zone if the designer can relate the $CO_2$ measurement in the return duct to breathing-level average measurements provided that same occupancy types and space usage are serviced by the return duct in that zone.
- In all rooms with $CO_2$ sensors, DCV controls should maintain $CO_2$ concentrations (with respect to the outdoor air $CO_2$ concentration) between the maximum level expected at design population and the minimum level expected at minimum population.
- Outdoor air $CO_2$ concentration should be measured continuously using a $CO_2$ sensor located in close proximity to the outdoor air intake. Alternatively, outdoor air $CO_2$ concentration can be assumed to be constant, provided the constant level is conservatively high and based on recent historical data for the area where the building is located. If an assumed value is used, consideration should be given in the controls to offsetting potential errors such as the tendency to overventilate at higher densities and underventilate at lower densities.
- $CO_2$ sensors should be specified by the manufacturer to have an uncertainty no greater than ±50 ppm for concentration ranges typically found in HVAC applications (e.g., 400 to 2000 ppm), be factory *and* field calibrated, and require calibration no more frequently than once every five years while operating under typical field conditions per manufacturer specifications (limited research indicates that field-based calibration should be performed once every one to two years [Fisk 2008]).
- Provisions (such as physical access and verification that the sensor is operating correctly) should be provided for periodic maintenance and calibration. This will assist in a) properly maintaining the DCV system and components and b) validating that the proper amount of ventilation is supplied under all variable occupancy levels and load conditions. Data logging of $CO_2$ concentrations can be considered; it allows review of $CO_2$ trend data in part to ensure that the $CO_2$ sensors and controls are operating as intended.

**Alternatives to $CO_2$-Based DCV**

In certain limited applications, such as classrooms, where occupancy is either zero or nearly 100%, the control of outdoor air rates in an on-off manner based on signals from a room occupancy sensor, time clock, or light switch is a practical and energy-saving solution. Other forms of DCV are based on technologies that can count the number of persons entering and exiting a space and adjust ventilation accordingly. In its simplest form this is done by estimating the number of persons during certain time periods and programming the ventilation supply accordingly. However, new advances in sensing and microcomputing technologies may automate this task. Dynamic infrared imaging hardware and software are now used for marketing and security purposes, and research proposals have been submitted to evaluate these technologies with DCV. New technologies have reduced signal delays and calibration drifts when compared to chemical-sensor-based DCV.

# $CO_2$-Based DCV Sensors

This case study illustrates some of the hardware used in applications of $CO_2$-based DCV. Figure 8.3-A shows a controller that can be used in single-zone constant-volume rooftop units. Figure 8.3-B depicts a conference room, which is a prime candidate for $CO_2$-based DCV given that the occupancy is variable and generally unpredictable. Figure 8.3-C shows how unobtrusive a wall-mounted $CO_2$ sensor can be.

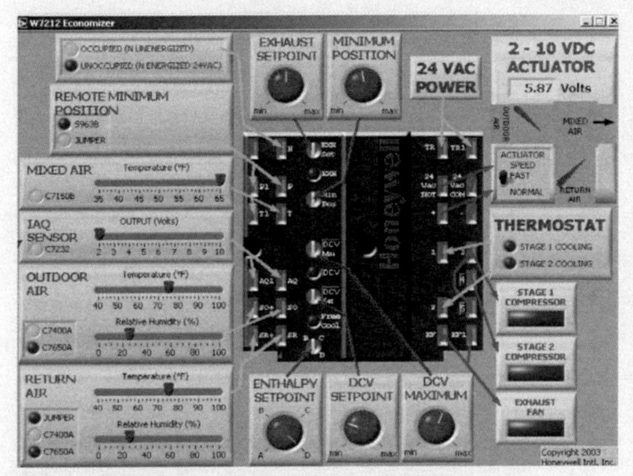


**Figure 8.3-A** Economizer Controller that Can Accept Reset Signal from a $CO_2$ Sensor

*Photograph copyright Honeywell International, Inc.*

**Figure 8.3-B** Wall-Mounted Sensor in Conference Room

*Photograph copyright GE Sensing & Inspection Technologies.*

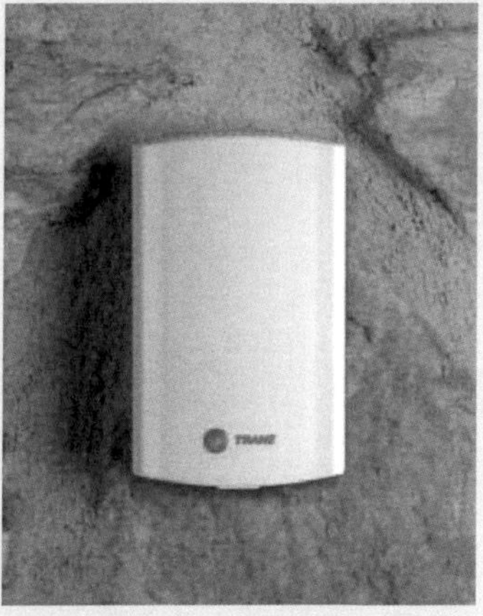

**Figure 8.3-C** Wall-Mounted Sensor

*Photograph copyright Trane.*

# Use Natural or Mixed-Mode Ventilation Where Appropriate



Natural ventilation has been used for thousands of years to ventilate and cool spaces. Many examples can be found in ancient building structures from the Pantheon in Rome, where the "Great Eye" at the dome's apex is the primary source of ventilation as well as daylight, to Persian building structures, which use wind scoops called *malqafs* and water features designed to take advantage of the wind, evaporative cooling, and natural buoyancy effects to passively ventilate and cool the building (Walker 2008).

Naturally ventilated buildings do not aim to achieve constant environmental conditions but do take advantage of, and adapt to, dynamic ambient conditions to provide a controllable, comfortable indoor environment for the occupants.

Clearly there are locations in North America that are not suitable for natural ventilation/natural cooling—especially where tight temperature and humidity control is required or in locations that experience prolonged periods of high outdoor temperature, high humidity, chronic outdoor air pollution, or other severe outside weather conditions. On the other hand, there are large portions of North America that can take advantage of natural ventilation strategies for the whole of the year or a significant portion of the year.

When considering the use of natural ventilation, early consideration/analysis of the appropriateness of the prevailing climate must be evaluated in some detail. Climatic issues such as the ambient air temperatures, humidity, and cleanliness of the outdoor air and wind airflow patterns need to be considered.

Natural ventilation and cooling generally works well with other sustainable strategies; for instance, energy-efficient design typically requires the reduction/control of thermal gains and losses, which in turn is an essential design component for natural ventilation. Daylit buildings with narrow floor plates and high floor-to-ceiling areas work well for natural ventilation. Also, naturally ventilated buildings often take advantage of thermally massive elements such as concrete/masonry structures to provide a more stable mean radiant temperature by absorbing and releasing heat slowly. Mass can also be used to temper incoming air, especially when using night flushing, and can moderate mean radiant temperature, which improves comfort. Mass provides a thermal damper, so the building requires less overall energy to heat and cool (Willmert 2001). Displacement ventilation and decoupled DOAS strategies can work well with natural ventilation, as well.

Natural/mixed-mode ventilation systems are now more common in large buildings, especially in the Pacific North West, Japan, and Europe. With mixed-mode systems, natural ventilation is commonly used for ventilating/cooling for most of the year and mechanical ventilation/cooling systems are used for peak cooling and when natural ventilation is not available.

Also, pressure sensors and motor-driven dampers are being used to control pressures in various parts of buildings and to take advantage of stack effect or wind pressure to deliver ventilation where and when it is needed. These sophisticated ventilation control systems need considerable care in design and operation as well as end-user education.

Key issues for consideration in selecting and designing natural ventilation systems include:

- delivering sufficient outdoor air to dilute indoor pollutants and maintain the required thermal comfort (in accordance with ASHRAE Standard 62.1 [ASHRAE 2007a] and ASHRAE Standard 55 [ASHRAE 2004]);

8.4

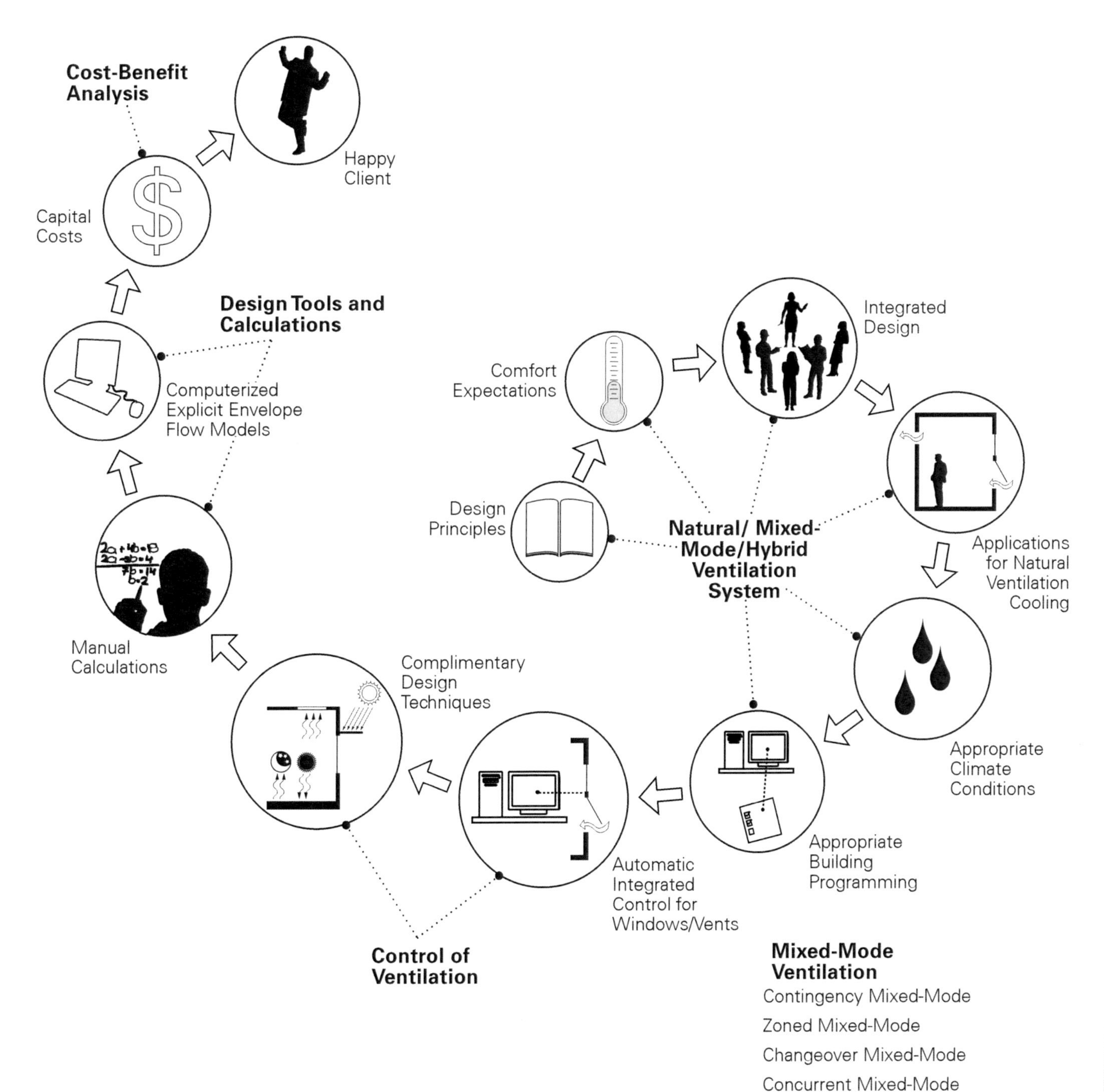


**Mixed-Mode Ventilation**

Contingency Mixed-Mode

Zoned Mixed-Mode

Changeover Mixed-Mode

Concurrent Mixed-Mode

- reducing the entry of undesirable constituents in polluted outdoor air;
- good solar control and modest internal gains plus an acceptance that the internal temperature will exceed 77°F (25°C) for some period of time (CIBSE 2005);
- controlling airflow through passive or active means, which requires well-designed systems with thorough consideration of airflows under the wide range of outdoor weather conditions to which the building will be subjected;
- a satisfactory acoustic environment (natural ventilation openings provide a noise transmission path from outside to inside, which may be a determining factor in some building locations; in addition, naturally ventilated buildings often include large areas of exposed concrete in order to increase the thermal capacity of the space, and such large areas of hard surface require careful attention to achieve a satisfactory acoustic environment for the occupants);
- smoke control (since smoke can follow natural ventilation paths, the integration of the fire safety strategy must be integrated with the natural ventilation design); and
- health and safety (many natural ventilation openings will be at significant heights above floor level, so safe/easy access to these openings/control devices is required to be considered in the design).

## The Carnegie Institute for Global Ecology

**Figure 8.4-A** Carnegie Institute for Global Ecology
*Photograph courtesy of Paul Sterbentz, Carnegie Institution of Washington.*

The Carnegie Institute for Global Ecology is an airy, daylit 11,000 ft2 (1022 m2) building on the Stanford University campus (Figure 8.4-A). The building was completed in 2004 by a project team including members from the architect firm Esherick Homsey Dodge and Davis (EHDD) and Rumsey Engineers and Engineering Enterprise. The designers predicted that the building would use 45% less energy and 40% less water than a code-compliant equivalent building. AIA recognized this extremely low-energy laboratory and office building as one of the Top Ten Green Projects of 2007. To control temperature, the building uses radiant cooling and operable windows in the upstairs office levels; only the laboratories are mechanically ventilated. An occupant survey conducted in the building indicated satisfaction from the users of the building in part due to the natural ventilation used in the design (Figure 8.4-B).

**Occupant Satisfaction**

highly dissatisfied / neutral / highly satisfied
-3 -2 -1 -0 1 2 3
General Building Satisfaction
Thermal Comfort
Air Quality
Lighting
Acoustics
◇ CBE Survey Average
○ Carnegie Center

**Figure 8.4-B** Occupant Satisfaction Chart
*Adapted from CBE (2005).*

*Case study source: Weeks et al. (2007) and CBE (2005).*

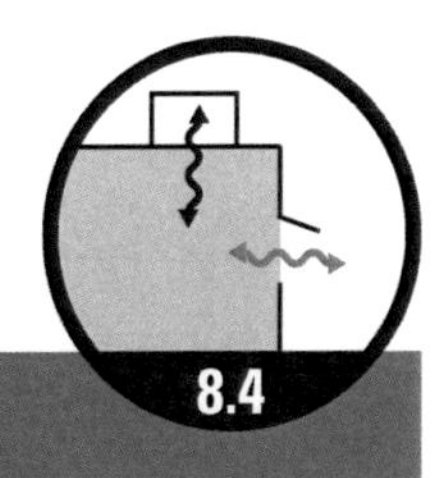


## Liberty Tower of Meiji University

Liberty Tower, the high-rise school building of Meiji University in Japan (Figure 8.4-C), is 570,506 $ft^2$ (53,000 $m^2$) and has 23 floors above grade. It was completed by a project team including members from Nikken Sekkei Ltd. and Toshiharu IKAGA. The building shape and design (Figure 8.4-D) were determined to maximize natural ventilation, with controlled windows and a wind floor (Figure 8.4-E) automatically controlling the building pressure and airflow. The building optimizes the use of natural lighting and incorporated a number of energy savings measures. The decrease in primary energy consumption is shown in Figure 8.4-F.

**Figure 8.4-C** Liberty Tower
*Photograph courtesy of Kato and Chikamoto (2002).*

**Figure 8.4-E** Wind Floor
*Photograph courtesy of Kato and Chikamoto (2002).*

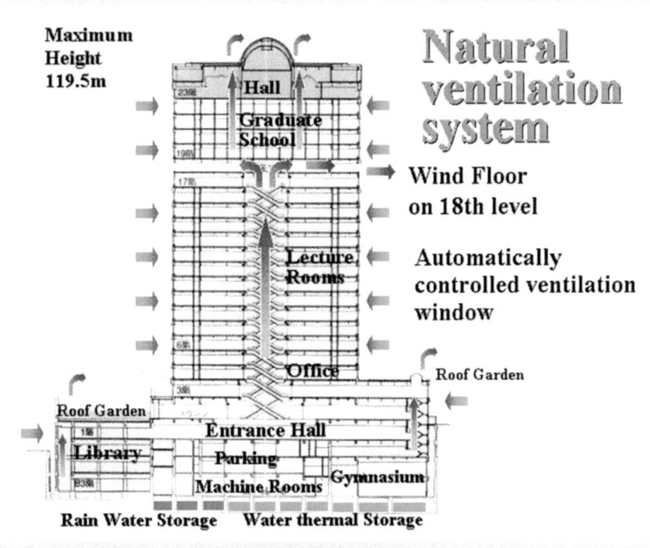


**Figure 8.4-D** Airflow Schematic for the Natural Ventilation System
*Image courtesy of Kato and Chikamoto (2002).*

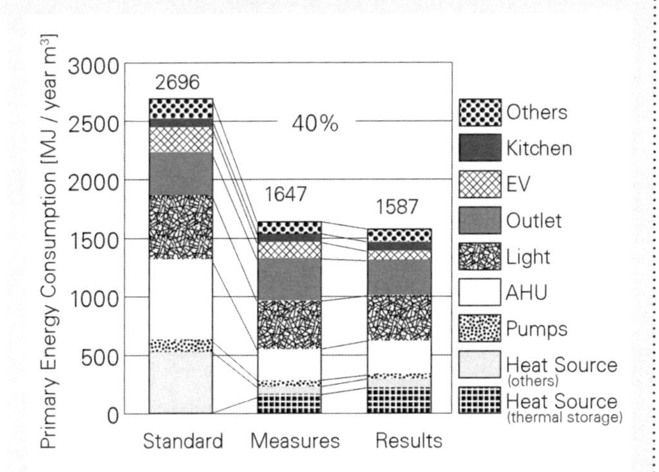


**Figure 8.4-F** Measured Energy Consumption
*Adapted from Kato and Chikamoto (2002).*

STRATEGY OBJECTIVE 8.4

# Use the ASHRAE Standard 62.1 IAQ Procedure Where Appropriate



The IAQ Procedure (IAQP) provides designers with an important option or adjunct to the prescriptive Ventilation Rate Procedure (VRP) in ASHRAE Standard 62.1 (ASHRAE 2007a), thereby increasing the potential for good IAQ control.

In general, the attainment of good IAQ can be achieved through the removal or control of irritating, harmful, and unpleasant constituents in the indoor environment. The established methods for contaminant control are source control, ventilation, and filtration and air cleaning. Source control approaches should always be explored and applied first because they are usually more cost-effective than either ventilation (dilution) or FAC (extraction). Other Strategies in this Guide discuss various aspects of ventilation for attainment of acceptable IAQ as presented by ASHRAE Standard 62.1-2007. However, the main focus of ASHRAE Standard 62.1 is the VRP, which specifies minimum outdoor air ventilation rates to dilute indoor contaminants.

This discussion focuses on an alternative compliance pathway to the VRP that is referred to in the standard as the *IAQ Procedure*. The application of the IAQP typically employs a combination of source control and enhanced extraction through filtration and/or gas-phase chemical air cleaning, in some cases resulting in a reduction in the minimum outdoor air intake flow required, compared to the more commonly used prescriptive VRP. (See Strategy 7.5 – Provide Particle Filtration and Gas-Phase Air Cleaning Consistent with Project IAQ Objectives).

The IAQP generally employs all three basic control methods—source control, ventilation, and FAC—allowing one to realize the combined strengths of each to yield the following potential benefits:

- A methodology for documenting and predicting the outcome of source control approaches and rewarding source reduction tactics by potentially lowering ventilation requirements.
- It can lower the heat, moisture, and pollutant burden of outdoor air by reducing the outdoor airflow rate to the conditioned space.
- The use of enhanced FAC lowers the constituent contaminant concentrations of CoC contained in the outdoor air.
- The use of enhanced FAC can lower the constituent concentration of CoC created and recirculated within the conditioned space.
- Enhanced FAC can result in cleaner heat exchange surfaces and more energy-efficient HVAC system operation.
- Lower outdoor air intake rates can lower system capacity and operating costs.

The IAQP (ASHRAE 2007a) was first introduced in the original ASHRAE Standard 62 in 1973, discussed in ASHRAE Standard 62-1981, and formalized in ASHRAE Standard 62-1989 as an alternate path of compliance to attain acceptable IAQ. However, the IAQP was not widely accepted by model code bodies, so approval by the authority having jurisdiction typically requires a code variance. For this reason, the procedure has not been widely used by designers, who are also reluctant to use it because of its complexity, the potential liability involved, and the additional engineering rigor required, including more calculations, analyses, and/or testing. The occasional use of the procedure has been predominately in areas having high outdoor humidity and heat loads; in buildings having high internal contaminant generation; and in buildings having high density and diversity, such as arenas, schools, auditoriums, theaters, convention centers, and hotels. In

IAQ 62.1 8.5

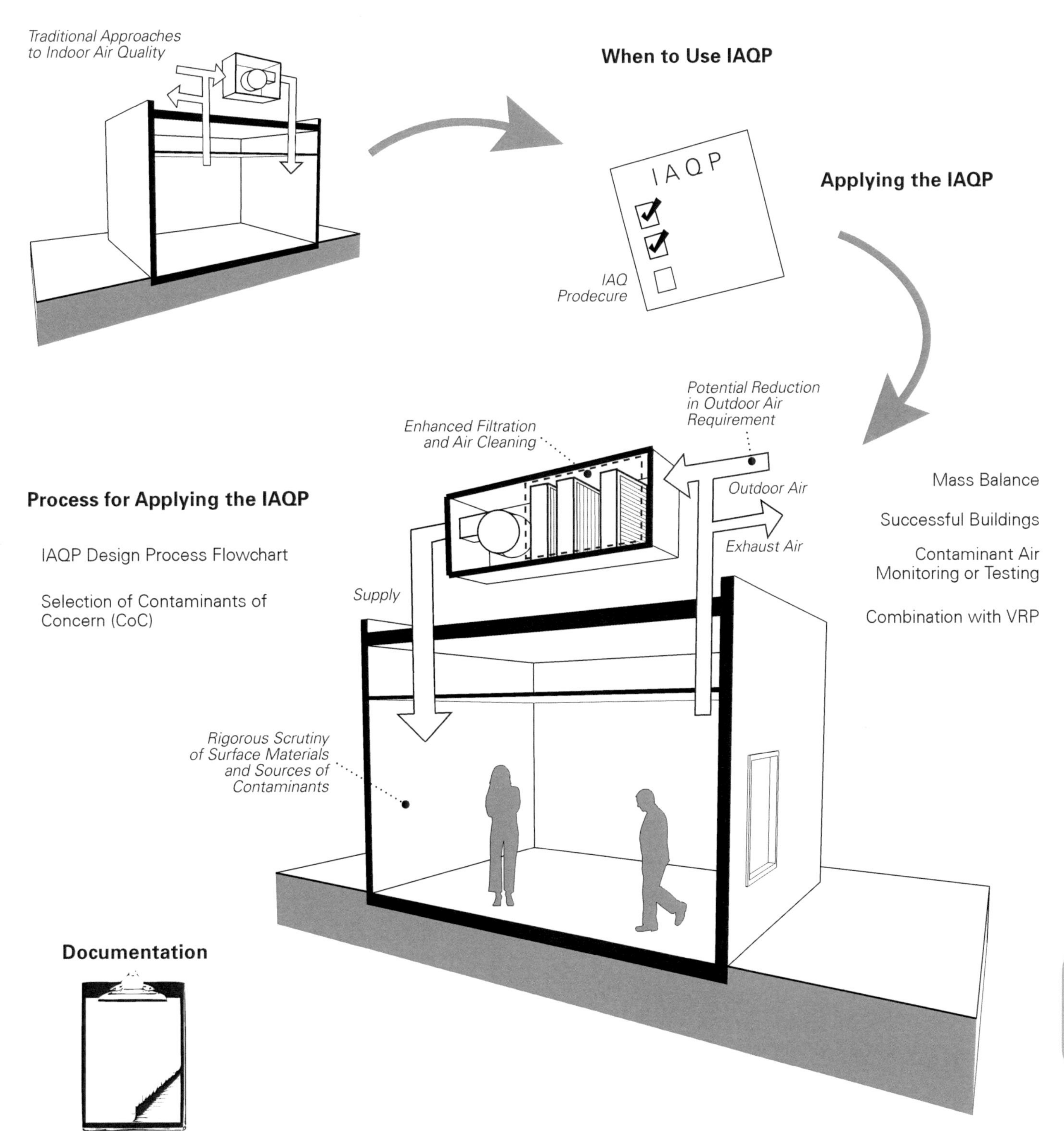
History of the IAQP
Traditional Approaches to Indoor Air Quality
When to Use IAQP
IAQP
IAQ Prodecure
Applying the IAQP
Potential Reduction in Outdoor Air Requirement
Enhanced Filtration and Air Cleaning
Outdoor Air
Exhaust Air
Supply
Mass Balance
Successful Buildings
Contaminant Air Monitoring or Testing
Combination with VRP
Process for Applying the IAQP
IAQP Design Process Flowchart
Selection of Contaminants of Concern (CoC)
Rigorous Scrutiny of Surface Materials and Sources of Contaminants
Documentation

these selected spaces, the economic advantages can provide compelling returns on the additional design and equipment investment, as illustrated by the case study titled "Using the IAQP in Building Design and Construction."

**Related Strategies**

There is considerable interaction of the application of the IAQP to other areas of the design process. These include the following:

- The evaluation of sources, components, and concentrations of CoC in the outdoor air. See the Strategies in Objective 3 – Limit Entry of Outdoor Contaminants.
- The selection and evaluation of the building component materials for their outgassing and contaminant source contributions. See the Strategies in Objective 5 – Limit Contaminants from Indoor Sources.
- The determination of filtration needs and the selection of the appropriate FAC efficacy and equipment type. See Strategy 7.5 – Provide Particle Filtration and Gas-Phase Air Cleaning Consistent with Project IAQ Objectives.
- The evaluation and selection of the ventilation system. See the Strategies in Objective 7 – Reduce Contaminant Concentrations through Ventilation, Filtration, and Air Cleaning.
- The selection of HVAC systems and equipment so that enhanced filtration can be installed with adequate access, space, and fan horsepower. See Strategy 1.3 – Select HVAC Systems to Improve IAQ and Reduce the Energy Impacts of Ventilation.

**Applying the IAQP**

ASHRAE Standard 62.1-2007 allows several alternative approaches of applying the IAQP. They include a mass balance approach using steady-state calculations of CoC, a comparison with similar buildings approach to document successful usage elsewhere, and a contaminant monitoring approach where actual contaminant levels are monitored. By following a set of predefined steps, it is possible to both enhance the IAQ and substantially reduce the outdoor airflow requirements in some buildings. The steps include evaluation of CoC, target levels of acceptability, methods for determining acceptability, examination of ventilation requirements, material selection, FAC options, and implementation and documentation of the IAQP. (See the Part II detailed guidance on this Strategy for further details on applying the IAQP.)

IAQ 62.1

8.5

## Using the IAQP in Building Design and Construction

A large public assembly building located in a major Southeastern city (Figure 8.5-A) was constructed in 1991 using the IAQP of ASHRAE Standard 62-1989, which was part of the applicable Southern Building Code at the time. Enhanced filtration was included in the original design because the facility is located in an urban area near dense auto, truck, and train traffic. The facility features public events that generate high levels of internal dust and dirt loads, such as monster truck and tractor pulling, that led to the need for enhanced filtration. This filtration component enabled the use of the IAQP, which reduced the peak outdoor air requirement two thirds, from 15 to 5 cfm (7.5 to 2.5 L/s) per person (or over 750,000 cfm [375,000 L/s] total). This reduction yielded a reduction of 2350 tons (8260 kW) of chiller capacity and 2.5 million dollars of related construction costs. Reduced operating energy costs totaled 40 million Btu (42 million kJ), or $800,000 per year in 1991 dollars, which equals $1,256,000 per year in 2008 dollars. This highly successful installation currently services their MERV 13 bag filters annually. Staff estimate the total accumulated savings to date to be from $13,000,000 to $15,000,000, which is more than the original cost of the building.

**Figure 8.5-A** Large Sports Arena in Southern United States
*Photograph courtesy of H.E. Burroughs.*

Performance verification conducted in 2006 revealed that the FAC system efficiency for particle reduction exceeded 90% at the 0.5 μm size. TVOC concentration of the supply air did not exceed 106 μg/m$^3$. No ozone was detectable downstream of the filter bank though outdoor challenge concentrations ranged from 11 to 48 ppb (22 to 94 μg/m$^3$).

## Part I References and Bibliography


AAMA. 1988. *Window Selection Guide*. Schaumburg, IL: American Architectural Manufacturers Association.

AIA. 2006. *Guidelines for Design and Construction of Health Care Facilities*. Washington, DC: American Institute of Architects.

Alevantis, L.E., R. Miller, H. Levin, and J.M. Waldman. 2006. Lessons learned from product testing, source evaluation, and air sampling from a five-building sustainable office complex. Proceedings Healthy Buildings 2006, Lisbon, Portugal, June 4–8.

Archiplanet. 2009. Philip Merrill Environmental Center. Category: Livable Buildings Awards 2007. Eugene, OR: Archiplanet. www.archiplanet.org/wiki/Philip_Merrill_Environmental_Center.

ASHRAE. 1997. *ASHRAE Thermal Comfort Tool*. Atlanta: American Society of Heating, Refrigerating and Air Conditioning Engineers, Inc.

ASHRAE. 2004. *ANSI/ASHRAE Standard 55, Thermal Environmental Conditions for Human Occupancy*. Atlanta: American Society of Heating, Refrigerating and Air-Conditioning Engineers, Inc.

ASHRAE. 2005. *ASHRAE Guideline 0-2005, The Commissioning Process*. Atlanta: American Society of Heating, Refrigerating and Air-Conditioning Engineers, Inc.

ASHRAE. 2007a. *ANSI/ASHRAE Standard 62.1-2007, Ventilation for Acceptable Indoor Air Quality*. Atlanta: American Society of Heating, Refrigerating and Air-Conditioning Engineers, Inc.

ASHRAE. 2007b. *ASHRAE Guideline 1.1-2007, HVAC&R Technical Requirements for The Commissioning Process*. Atlanta: American Society of Heating, Refrigerating and Air-Conditioning Engineers, Inc.

ASHRAE. 2007c. *ANSI/ASHRAE Standard 52.2-2007, Method of Testing General Ventilation Air-Cleaning Devices for Removal Efficiency by Particle Size*. Atlanta: American Society of Heating, Refrigerating and Air-Conditioning Engineers, Inc.

ASHRAE. 2007d. *62.1 User's Manual*. Atlanta: American Society of Heating, Refrigerating and Air-Conditioning Engineers, Inc.

ASHRAE. 2007e. *ANSI/ASHRAE/IESNA Standard 90.1, Energy Standard for Buildings Except Low-Rise Residential Buildings*. Atlanta: American Society of Heating, Refrigerating and Air-Conditioning Engineers, Inc.

ASHRAE. 2008a. *ASHRAE Position Document on Environmental Tobacco Smoke*. Atlanta: American Society of Heating, Refrigerating and Air-Conditioning Engineers, Inc.

ASHRAE. 2008b. *ANSI/ASHRAE/ACCA Standard 180, Standard Practice for Inspection and Maintenance of Commercial Building HVAC Systems*. Atlanta: American Society of Heating, Refrigerating and Air-Conditioning Engineers, Inc.

ASHRAE. 2008c. *2008 ASHRAE Handbook—HVAC Systems and Equipment*. Atlanta: American Society of Heating, Refrigerating and Air-Conditioning Engineers, Inc.

ASHRAE. 2009. *2009 ASHRAE Handbook—Fundamentals*. Atlanta: American Society of Heating, Refrigerating and Air-Conditioning Engineers, Inc.

ASHVE. 1946. *1946 Heating Ventilating Air Conditioning Guide*. New York: American Society of Heating and Ventilating Engineers, pp. 184, 190.

ASTM. 2008. *ASTM E2600-08, Standard Practice for Assessment of Vapor Intrusion into Structures on Property Involved in Real Estate Transactions*. West Conshohocken, PA: ASTM International.

CBE. 2005. The Carnegie Institute for Global Ecology. Mixed Mode: Case Studies and Project Database. Berkeley, CA: Center for the Built Environment. www.cbe.berkeley.edu/mixedmode/carnegie.html.

CBE. 2007. Livable Buildings Awards. Berkeley, CA: Center for the Built Environment. www.cbe.berkeley.edu/livablebuildings/index.htm.

Chikamoto, T., S. Kato, T. Ikaga. 1999. Hybrid air-conditioning system at Liberty Tower of Meiji University. First International One Day Forum on Natural and Hybrid Ventilation, HybVent Forum'99, Sydney, Australia, September. http://hybvent.civil.auc.dk/puplications/report/chikamoto_et_al.pdf.

CIBSE. 2005. *AM10: Natural ventilation in non-domestic buildings*. London: Chartered Institution of Building Services Engineers.

CIWMB. 2000. *Section 01350, Special Environmental Requirements Specification*. Sacramento: California Integrated Waste Management Board. www.ciwmb.ca.gov/GreenBuilding/Specs/Section01350/.

CMHC. 1998. *Best Practice Guide: Building Technology Wood-Frame Envelopes in the Coastal Climate of British Columbia*. Ottawa, Ontario: Canada Mortgage and Housing Corporation.

DGS. 2004. About the project. East End Project. West Sacramento: California Department of General Services. www.eastend.dgs.ca.gov/AboutTheProject/default.htm.

Emmerich, S., and A.K. Persily. 2001. State-of-the-art review of $CO_2$ demand controlled ventilation technology and application. NISTIR 6729, National Institute of Standards and Technology, Gaithersburg, MD. Available at www.fire.nist.gov/bfrlpubs/build01/PDF/b01117.pdf.

DGS. 2004. About the project. DGC Real Estate Services. West Sacramento: California Department of General Services. www.eastend.dgs.ca.gov/AboutTheProject/default.htm.

EPA. 1989. Report to Congress on indoor air quality, volume II: Assessment and control of indoor air pollution. Report EPA/400/1-89/0001C, Office of Air and Radiation, U.S. Environmental Protection Agency, Washington, DC, August.

EPA. 2002. OSWER Draft Guidance for Evaluating the Vapor Intrusion to Indoor Air Pathway from Groundwater and Soils (Subsurface Vapor Intrusion Guidance). Washington, DC: U.S. Environmental Protection Agency, Office of Solid Waste and Emergency Response. www.epa.gov/correctiveaction/eis/vapor/complete.pdf.

EPA. 2007. Healthy School Environments Assessment Tool, Ver. 2 (HealthySEAT2). EPA IAQ Tools for Schools Program. Washington, DC: U.S. Environmental Protection Agency. www.epa.gov/schools/healthyseat/index.html.

EPA. 2008a. Indoor Air Quality Building Education and Assessment Model (I-BEAM). Washington, DC: U.S. Environmental Protection Agency. www.epa.gov/iaq/largebldgs/i-beam/index.html.

EPA. 2008b. National Ambient Air Quality Standards. Washington, DC: U.S. Environmental Protection Agency. www.epa.gov/air/criteria.html

EPA. 2008c. The Green Book Nonattainment Areas for Criteria Pollutants. Washington, DC: U.S. Environmental Protection Agency. www.epa.gov/air/oaqps/greenbk.

EPA. 2008d. Building Assessment Survey and Evaluation (BASE) Study. Washington, DC: U.S. Environmental Protection Agency. www.epa.gov/iaq/base/voc_master_list.html.

EPA. 2009. A Citizen's Guide to Radon. EPA 402/K-09/001, U.S. Environmental Protection Agency, Washington, DC.

Fischer, J.C., and C.W. Bayer .2003. Report card on humidity control—Failing grade for many schools. *ASHRAE Journal* 45(5):30–37.

Fisk, W.J. 2000. Health and productivity gains from better indoor environments and their relationship with building energy efficiency. *Annual Review of Energy and the Environment* 25(1): 537–66.

Fisk, W.J. 2008. Personal communication with Leon Alevantis. June 2007–November 2008.

Fisk, W.J., D. Faulkner, and D.P. Sullivan. 2005. An evaluation of three commercially available technologies for real-time measurement of rates of outdoor airflow into HVAC systems. *ASHRAE Transactions* 111(2):443–55.

Fisk, W.J., D. Faulkner, and D.P. Sullivan. 2007. A pilot study of the accuracy of $CO_2$ sensors in commercial buildings. *Proceedings of the IAQ 2007 Healthy and Sustainable Buildings, Baltimore, MD.*

Fisk, W.J., D. Sullivan, S. Cohen, and H. Han. 2008. Measuring outdoor air intake rates using electronic velocity sensors at louvers and downstream of airflow straightners. Report 1250, Lawrence Berkeley National Laboratory, Berkeley, CA.

Folkes, D.J. 2008. Principal and President, EnviroGroup Limited. Telephone interview with the author, January 7.

Franz, S.C. 1988. Architecture and commensal vertebrate pest management. *Architectural Design and Indoor Microbial Pollution*, ed. Ruth B. Kundsin. Oxford: Oxford University Press.

GINPR. 2009. Gulf Islands National Park Reserve of Canada. Parks Canada—Going Green. www.pc.gc.ca/pn-np/bc/gulf/ne/ne5_e.asp.

Heerwagen, J., and L. Zagrreus. 2005. The human factors of sustainable building design: Post-occupancy evaluation of the Philip Merrill Environmental Center, Annapolis, MD. Report prepared for the U.S. Department of Energy Building Technology Program, April. www.cbe.berkeley.edu/research/pdf_files/SR_CBF_2005.pdf

IAPMO. 2006a. Uniform Mechanical Code. Ontario, CA: International Association of Plumbing and Mechanical Officials.

IAPMO. 2006b. Uniform Plumbing Code. Ontario, CA: International Association of Plumbing and Mechanical Officials.

ICC. 2006a. 2006 International Mechanical Code. Country Club Hills, IL: International Code Council.

ICC. 2006b. 2006 International Plumbing Code. Country Club Hills, IL: International Code Council.

IEA. 1991. IEA Annex 14, Condensation and Energy. Final report, Vols. 1–4. Leuven: Acco Uitgeverij.

IOM. 2004. *Damp Indoor Spaces and Health.* Washington, DC: Institute of Medicine, National Academy of Sciences, National Academy Press.

IISBE. 2000. Liberty Tower of Meiji University – Case Study 4 – School. Selected GBC 2000 Case Studies. Green Building Challenge 2000. International Initiative for a Sustainable Built Environment. http://greenbuilding.ca/gbc2k/teams/Japan/Meiji/meiji-school.htm.

Int-Hout, D. 2007. Overhead heating: Revisiting a lost art. ASHRAE Journal 49(3):56–61.

ITRC. 2007. *Vapor Intrusion Pathway: A Practical Guideline.* VI-1. Washington, DC: Interstate Technology & Regulatory Council, Vapor Intrusion Team.

Kato, S., and T. Chikamoto. 2002. Pilot study report: The Liberty Tower of Meiji University, Tokyo, Japan. IEA ECBCS Annex 35: HybVent. http://hybvent.civil.auc.dk/pilot_study_buildings/Case_studies_pdf/CS7%20Liberty.pdf.

Latta, J.K. 1962. Water and building materials. Canadian Building Digest, CBD-30. www.nrc-cnrc.gc.ca/eng/ibp/irc/cbd/building-digest-30.html.

LBNL. 2008. *THERM (Two-Dimensional Building Heat-Transfer Modeling),* ver. 5.2. Berkeley, CA: Lawrence Berkeley National Laboratories.

LBNL. 2009a. WINDOW, ver. 5.2. Berkeley, CA: Lawrence Berkeley National Laboratories. http://windows.lbl.gov/software/window/window.html.

LBNL. 2009b. IAQ Scientific Findings Resource Bank (IAQ-SFRB), http:/eetd.lbl.gov/ied/sfrb/sfrb.html. Berkeley, CA: Lawrence Berkeley National Laboratories.

Lowe, J., J. Raphael, L. Lund, and R. Casselberry, Jr. 2009. Accelerating the redevelopment of a vapor-impacted property based on data-informed verification of vapor barrier technology. Air and Waste Management Association Vapor Intrusion 2009 Conference, San Diego, CA,

January 28–30.

Martin, S.B., C. Dunn, J. Freihaut, W.P. Bahnfleth, J. Lau, and A. Nedeljkovic-Davidovic. 2008. Ultraviolet germicidal irradiation, current best practices. *ASHRAE Journal* 50(8):28–36.

Mendell, M.J., and G.A. Heath. 2005. Do indoor pollutants and thermal conditions in schools influence student performance? A critical review of the literature. *Indoor Air* 15:27–52.

Mendell, M.J., W.J. Fisk, K. Kreiss, H. Levin, D. Alexander, W.S. Cain, J.R. Girman, C.J. Hines, P.A. Jensen, D.K. Milton, L.P. Rexroat, and K.M. Wallingford. 2002. Improving the health of workers in indoor environments: Priority research needs for a national occupational research agenda. *Am J Public Health* 92:1430–40.

Merchant, M. 2009. Professor and an Extension Urban Entomologist, Texas AgriLife Extension Service, Texas A&M, Dallas, TX. Personal communication with Mary Sue Lobenstein, January.

Milam, J. 1993. A holistic approach to improving indoor environmental quality. In *Designing Healthy Buildings—Paper Presentations.* Washington, DC: American Institute of Architects.

Milam, J.A. 1994. A holistic approach to improving indoor environmental quality. Strategic Planning for Energy and the Environment 13(4):19–35.

NAS. 1999. *Health Effects of Exposure to Radon: BEIR VI.* Washington, DC: National Academy of Sciences, National Academy Press.

NIBS. 2006. *NIBS Guideline 3-2006, Exterior Enclosure Technical Requirements for the Commissioning Process.* Washington, DC: National Institute of Building Sciences.

NY. 1999. High Performance Building Guidelines. New York: Department of Design and Construction. www.nyc.gov/html/ddc/downloads/pdf/guidelines.pdf.

OAQPS. 2009. A guide to air quality and your health. Air Quality Index. Washington, DC: U.S. Environmental Protection Agency, Office of Air Quality Planning and Standards, AIRNow. http://airnow.gov/index.cfm?action=aqibroch.aqi#9.

PCT. 2008. Experts discuss how to marry IPM with green-building design. Press release. Pest Control Technology. www.pctonline.com/articles/article.asp?ID=3241.

Perkins, J.E., A. Bahlke, and H. Silverman. 1947. Effect of ultra-violet irradiation of classrooms on spread of measles in large rural central schools. *American Journal of Public Health* 37:529–37.

Persily, A. 1997. Evaluating building IAQ and ventilation with indoor carbon dioxide. *ASHRAE Transactions* 102(2).

Schoen, L. 2006. Humidification reconsidered. *IAQ Applications* 7(1).

Schoen, L., R. Magee, and L. Alevantis. 2008. Construction with low emissions building materials and furnishings. Proc. Indoor Air 2008, 11th Int. Conf. on Indoor Air Quality and Climate, Copenhagen, August 17–22.

Seppanen, O., and W.J. Fisk. 2004. Summary of human responses to ventilation. *Indoor Air* 14(Supplement 7):102–18.

Shendell, D.G., R. Prill, W.J. Fisk, M.G. Apte, D. Blake, and D. Faulkner. 2004. Associations between classroom $CO_2$ concentrations and student attendance in Washington and Idaho. *Indoor Air* 14:333–41.

Shrestha, S., and G. Maxwell. 2009a. An experimental evaluation of HVAC-grade carbon dioxide sensors—Part I: Test and evaluation procedures. *ASHRAE Transactions* 115(2):471–483.

Shrestha, S., and G. Maxwell. 2009b. Wall-mounted carbon dioxide ($CO_2$) transmitters. National Building Controls Information Program, Iowa Energy Center, June. Available at www.energy.iastate.edu/Efficiency/Commercial/download_nbcip/PTR_CO2.pdf.

Shrestha, S., and G. Maxwell. 2010. An experimental evaluation of HVAC-grade carbon dioxide sensors—Part 2: Performance test results. *ASHRAE Transactions* 116(1).

SMACNA. 2005. *HVAC Duct Construction Standards; Metal & Flexible*, 3d ed. Chantilly, VA: Sheet Metal and Air Conditioning Contractors National Association.

Tillman, F.D., and J.W. Weaver. 2005. Review of Recent Research on Vapor Intrusion. EPA/600/R-05/106. Washington, DC: U.S. Environmental Protection Agency, Office of Research and Development.

Walker, A. 2008. The fundamentals of natural ventilation. Buildings, vol. 102 (October). www.buildings.com/Magazine/ArticleDetails/tabid/3413/ArticleID/6587/Default.aspx.

Wargocki, P., and D.P. Wyon. 2006. Effects of HVAC on student performance. *ASHRAE Journal* 48(10):22–28.

Weeks, K., D. Lehrer, and J. Bean. 2007. A model success: The Carnegie Institute for Global Ecology. Berkeley, CA: Center for the Built Environment (CBE). www.cbe.berkeley.edu/research/pdf_files/Weeks2007-CarnegieCaseStudy.pdf.

Wells, W.F. 1943. Air disinfection in day schools. *American Journal of Public Health* 33:1436–43.

Willmert, T. 2001. The return of natural ventilation. Architectural Record, Issue 7: 137–40, 142, 144–46. http://archrecord.construction.com/features/green/archives/0107ventilation-2.asp.

Wyon, D.P. 1996. Indoor environmental effects on productivity. *IAQ '96 Proceedings*, pp. 5–15.

Yuill, D.P., G.K. Yuill, and A.H. Coward. 2007. A study of multiple space effects on ventilation system efficiency in Standard 62.1-2004 and experimental validation of the multiple spaces equation. Final Report for ASHRAE Research Project 1276, American Society of Heating, Refrigerating and Air-Conditioning Engineers, Inc., Atlanta.